QUEEN REARING

Yours very truly,
L. E. Snelgrove.

Frontispiece

QUEEN REARING

By

L. E. SNELGROVE, M.A., M.Sc.

Fellow of the Royal Entomological Society
President of the Somerset Beekeepers' Association
A Vice-President of the British Beekeepers' Association
Expert and Honours Lecturer, British Beekeepers' Association
Member of the Apis Club

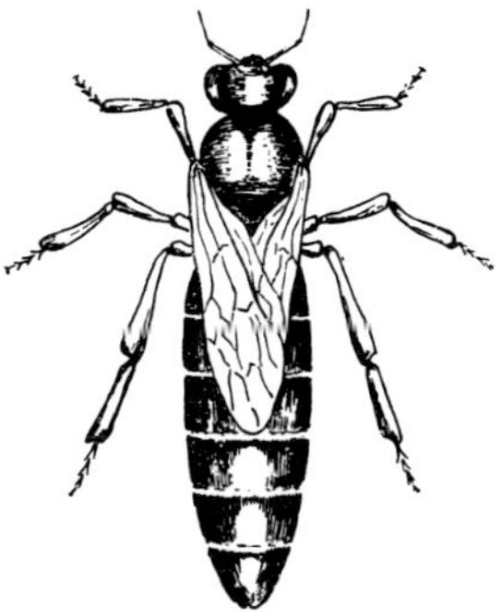

Fourth Edition *June 1981*

Published by

SNELGROVE & SMITH
PLEASANT VIEW, BLEADON HILL, WESTON-SUPER-MARE
AVON BS24 9JT

MADE AND PRINTED IN GREAT BRITAIN BY PURNELL AND SONS, LTD.
PAULTON (BRISTOL) AND LONDON

INTRODUCTION

THE value of bees in relation to a nation's food supply is being increasingly recognised. As honey producers they provide a delectable and wholesome sweet and the wax they secrete is inimitable and unexcelled. It is however in their work of pollination that their chief value lies for not only are they mainly responsible for the production of good fruit crops but their agency is necessary for the seeding and spread of some of the principal forage crops on the farm. .

During the recent great war there has been a notable extension of beekeeping in most countries, due largely to help given by the respective governments. This help has usually taken the form of encouraging more people to keep bees and giving instruction in management. There has however been little or no general effort to increase the productiveness of bees by the improvement of stock, as has been made, for example, in the case of British cattle, although a considerable amount of scientific work is now being done with this end in view under the aegis of the U.S. Department of Agriculture, and at the British Ministry of Agriculture's Experimental Station at Rothamstead. Beekeepers themselves, whilst aware of the need for improvement, usually leave selection to nature with the result that a large proportion of the stocks in the countryside are relatively unproductive.

A few years ago the beekeeping industry in Britain stood in need of improvements in respect of management, the control of diseases, the cultivation of nectar-bearing flora, and the selection of stock. Since then improved systems of management have deen devised, effective treatments of the more serious diseases brought into use, and pastures enriched by the introduction of new nectar-bearing clovers. In the field which offers by far the greatest scope for progress however, namely, that of stock improvement, little or no concerted effort has been made or even encouraged. It is true that the importation of cheap foreign queens has been suspended and that it is unlikely that they will again be obtained at absurdly low prices. There is good reason therefore why scientific queen-rearing should be widely developed in the country and British beekeepers become more self-supporting for their supply of queens.

The efforts of reliable queen-breeders and of associations owning breeding stations can effect something in the direction of improvement, but owing to the uncontrolled mating of bees, the existence of a few inferior stocks in a particular neighbourhood will tend to neutralise any improvement made. It is therefore the indifferent small-scale beekeeper who should be made aware of the advantages of selective breeding and who should be encouraged to breed and maintain superior stock.

The scientific rearing of queens from choice stock is acknowledged to be one of the most difficult aspects of beekeeping and many people regard it as beyond their capabilities. There are however some very simple methods which can be improved and made effective and there is no reason why even the smallest-scale beekeeper should not make his contribution to a general improvement.

At the present time there is no book in circulation which deals comprehensively and exclusively with the subject of queen-rearing. Doolittle's pioneer work, first published nearly sixty years ago, remains unrevised. The works of Sladen (England), and Perret-Maisonneuve (France), are out of print, as is also, at least for the time being, Herrod-Hempsall's encyclopaedic work "Beekeeping New and Old" which contains a full and well illustrated description of queen-rearing practice.

Some of the more important works on general beekeeping, notably Root's ABC and XYZ of Bee Culture (America), Wedmore's "A Manual of Beekeeping", and Manley's "Honey Production in the British Isles" contain authoritative and well written sections dealing with the subject, and many excellent articles relating to it appear as bulletins of educational bodies and contributions to the beekeeping journals.

Jay Smith's "Queen Rearing Simplified" deals admirably with the practical side of the subject as followed in America, but no English book specifically devoted to Queen Rearing has appeared since the second edition of Sladen's excellent little work "Queen Rearing in England" was last published in 1913.

I regret that Mr. F. C. Pellett's "Practical Queen Rearing" (America), which has been described as a clear exposition of the principal methods of queen-rearing, has not yet come to my notice.

In compliance therefore with requests from several prominent beekeepers I have endeavoured not only to fill an obvious void in British beekeeping literature, but also, by including much that is new both in theory and practice, to bring the whole subject up-to-date. I trust that the book will serve the small-scale beekeeper and the large-scale beekeeper equally well, and at the same time be of some service to the professional queen-breeder. I hope too that it will be useful to students who enter for advanced examinations in apiculture.

For the convenience of readers I have divided the book into two parts, the first dealing with what may be considered the scientific background of the subject and the second with the practical side. Preliminary study of the first part is essential if the fullest advantage is to be obtained from the application of the second.

In dealing with the difficult subject of heredity in bees I have attempted to clear up some popular misconceptions concerning "Mendelism" in its relation to the breeding of bees. I have also dealt at some length with the subject of larval nutrition which is of prime importance in the raising of queens.

In the practical section some of the directions I have given may appear to experienced workers to be unnecessarily detailed. My aim has been to prevent the less experienced from making mistakes which so often arise from following abbreviated instructions.

With regard to the illustrations every necessary manipulation and every essential piece of equipment is depicted. The latter is of the simplest type and such as I have found both adequate and satisfactory in use. I have not felt justified in increasing the size of the book by describing alternative articles which have been invented and used by others but which have no superior value.

Most of the illustrations are from photographs taken by myself in my own apiaries. My assistants therefore appear in manipulations.

I have not attempted to depict exceptionally large acceptances of artificial queen-cells. These could easily have been obtained by the simple expedient of using very large quantities of bees in the cell-raising stocks. Whilst attractive in illustrations they are not necessarily satisfactory and they lead the inexperienced breeder to expect much better results than are ordinarily obtainable. The results shown in this book include successes and failures such as may be expected from the use of the bees carried by a single British Standard brood chamber, and likely to be attained by any careful manipulator.

References to other illustrations are made below.

It now remains for me to acknowledge the valuable help which I have received from various sources.

In the first place I am greatly indebted to the following for scientific papers and helpful personal communications:

(1) Professor M. K. Haydak of the University of Minnesota who sent me several papers describing the work of himself and his collaborators in the study of bee nutrition, and particularly his comprehensive paper on "Larval Food and the Development of Castes in the Honeybee".

(2) Dr. Otto Mackensen of the State University of Louisiana, who sent me his paper on "The Occurrence of Parthenogenetic Females in some Strains of Honeybees".

(3) Mr. P. B. Pearson of the Agricultural and Mechanical College of Texas for papers on The Pantothenic Acid Content of Royal Jelly (Pearson and Burgin), and of Pollen, (Pearson).

(4) Dr. C. G Butler of the Rothamstead Experimental Station, Herts, for papers relating to Brood Diseases of Bees and his Surveys of Adult Bee Diseases in England and Wales.

(5) Mr. L. Illingworth, B.A., Secretary of the Apis Club.

I have also derived considerable help from Miss A. D. Betts' writings and reviews of foreign scientific papers in "The Bee World".

I am further indebted to Professor Harris of the University of Bristol who kindly gave me facilities for consultations in the department of Zoology of the University.

But chiefly am I beholden to Mr. R. Bassindale, M.Sc., lecturer in Zoology in the University of Bristol. Not only has he read the whole of the typescript and made many helpful criticisms and suggestions, but he has, in addition, from material provided by me, made a number of anatomical preparations of the generative systems of the queen and drone, and the unique photomicrographs of them which appear in Plates II, III, VI, XI, and XV.

For other illustrations not my own I am indebted to

(1) The McGraw Hill Book Company, New York, for permission to reproduce Snodgrass' well-known diagram of the generative organs, etc., of the queen-bee (Plate V).

(2) Mr. J. F. Bramwell for the photograph in Plate XIX.

(3) Mrs. M. Hewison for photographs by the late Rev. G. H. Hewison (Plates VIII and XVII).

(4) Mr. H. Teal for the photograph in Plate IV.

The diagram in Plate XIII was drawn for me by Mr. E. G. Davies, and the drawings in the title page and in Plate XXIV were made by Miss I. Snelgrove.

I must acknowledge too, the valuable help I have received in practical queen-rearing from my assistant Mr. R. C. Cook, and also the ever willing assistance of Mr. J. Richards. Both are skilled wood-workers and have made whatever apparatus I have needed.

My thanks are due to Mr. J. Spiller and Mr. A. G. Pilkington, both of whom have experience of queen-rearing, for kindly reading the proofs before publication.

For other incidental help I am grateful to Miss M. Bindley, Miss I. Snelgrove, and Mrs. K. Horsey.

The works included in the bibliography at the end of the book are numbered, and references to them are made in the text. In each case the number of the work is followed by the number of the page to which reference is made.

L. E. SNELGROVE.

Bleadon,

April, 1946

CONTENTS

LIST OF ILLUSTRATIONS

PART I

CHAPTER I

IMPORTANCE OF QUEEN-REARING

Every experienced beekeeper knows that the prosperity and productiveness of a stock of bees depend primarily on the age and qualities of its queen. He is aware, too, that other things being equal, young queens are more prolific than old ones, and that there may be marked variations in the appearance, temper, industry, longevity, and disease-resisting qualities of their progeny. He therefore desires that his stocks be headed by queens having the most desirable characteristics, and in every season is confronted with the problem of raising, or otherwise providing them.

Scientific queen-rearing however is generally regarded as the most difficult branch of apiculture. It demands sound knowledge of the natural history of the bee, manipulative skill, the use of specialised equipment, and the expenditure of much time during and after the honey season. For these reasons it is much neglected by small-scale beekeepers. It is mainly practised in this country by a limited number of breeders who find a ready market for their queens.

The small-scale beekeeper often leaves re-queening to nature—with consequent losses which will be referred to in a later chapter—or purchases young queens from British or foreign breeders who may be more or less reliable. Indeed some reputable writers on beekeeping go so far as to advise the small bee-keeper to purchase his queens rather than to rear his own, on the ground that, although it is easy, it is in the long run unprofitable to raise queens unskilfully and of inferior quality.

This is true, not only of bees but of all useful animals and plants. For example we are all familiar with the farmer who appears to be content with cows which yield from 300 to 400 gallons of milk per year. His neighbour, starting with a good strain, and by constant and careful selection, secures cows giving from 800 to 1,000 gallons per year, whilst the record yield from one cow, the result of long continued artificial selection, has exceeded 3,000 gallons. The wild jungle fowl from which our domestic fowl has descended, lays one or two clutches of eggs in a year. Her modern descendant, partly on account of changed environment, but mainly of continued selective breeding, may produce well over 300 eggs in her first laying year. Similarly, horses are improved for speed and strength, sheep for meat and wool, and indeed, all domesticated animals for the particular qualities considered desirable in them.

The high-yielding varieties of corn we grow, the vegetables we eat, and the marvellous variety of beautiful flowers which adorn our greenhouses and gardens have descended from wild and humble ancestors and are constantly being improved by selection and crossing.

It is much the same in the case of bees. We may select for utility or appearance. Forty years ago there was a popular desire for highly coloured bees, e.g. "five-banded goldens". These were produced and distributed but at the expense of stamina and productivity. At the present time appearance is not a primary consideration except in so far as it indicates descent. Of much greater importance are such qualities as prolificacy, longevity, industry, disposition, tendency to abstain from swarming, and to store honey in excess of winter needs.

The modern hive, with its capacious brood chamber —often relieved or enlarged by the beekeeper—tends to exhaust the laying powers of a queen so that it is usually unprofitable to keep her after two full seasons. As the beekeeper desires to have the largest possible stocks for the nectar gathering he should renew his queens at least every other year, and having due regard for other characteristics, should breed from those which are the most productive. By careful and intelligent selection he may develop a strain of his own which will gradually increase his profits and at the same time add greatly to the interest and pleasure he derives from his beekeeping.

Young fertile queens are needed:—

(1) For annual or biennial requeening.
(2) For increase.
(3) As substitutes for failing or defective queens.
(4) When a queen has been lost by accident or has died of disease.
(5) When young queens have failed to mate or are lost on their mating flights.
(6) When a fertile queen has been lost on a cleansing flight which some queens are believed to take in the spring.
(7) As surplus queens for wintering.

In most countries there are beekeepers who specialise in the breeding of queens for market. In this country queen-rearing is usually a side-line to honey production.

Young queens are unobtainable before May but are fairly plentiful from July to September. At the time of writing they are expensive, varying in price from ten shillings to thirty shillings each, and it by no means follows that the higher-priced ones are better than the others. A great deal of money is wasted every year in the purchase of indifferently bred queens. Indeed, the present writer, after forty years of experience with bees, during the first thirty of which he bought foreign queens rather freely, has rarely been so fortunate as to purchase one of outstanding excellence.

The question then arises,—can the small-scale beekeeper, whom we may consider as one possessing from one to forty hives, reasonably hope to produce queens of good quality and at the same time work for honey production? There are some writers who would answer the question with a definite "No." They say that queen-rearing should be left to the specialist who has the necessary time, skill, and facilities for rearing under the best conditions, and who is able systematically to select both queens and drones for breeding.

There is no reason, however, why the amateur who has had two or three years' experience with bees, has the necessary time and a little additional equipment, and is prepared to observe a few cardinal principles, should not raise excellent queens, gradually improve his stock, and materially increase his production of honey.

It will be the aim of the writer, by means of simple explanations and instructions, to show how this can be done.

CHAPTER II

QUEEN-REARING IN NATURE

LET us first consider in some detail how bees in a state of nature raise their queens. It is only by a full understanding of nature's ways that we can appreciate the necessity of observing the various times, conditions, and rules essential for success in rearing queens artificially.

There are some who contend that naturally-raised queens are the best, and in a general sense this may be true. We can however improve on nature in respect of selection and the number of queens obtainable, and we can within a season's limits breed them at times of our own choosing.

For convenience we will consider the life of a young queen as from the beginning of a new year. She was bred during the preceding summer and so in January will be six or seven months old.

Since October she and her bees have clustered in a more or less compact mass, eating only sufficient food to maintain the temperature requisite for existence. Shortly after the turn of the days—sometimes later—the bees begin to consume a little more food, the temperature of the cluster rises, and the queen, receiving stimulating food from the workers, deposits her first few eggs of the season in cells already cleared and cleaned by the bees. A spell of cold weather may retard this early development, but mild weather will promote it and the bees will be stirred into increasing activity. They carry water and pollen and search for nectar in the early blossoms. The brood nest and the population slowly increase until the hive is filled with brood, pollen, honey, and bees.

Before this state is reached however, the queen lays a few eggs in drone cells. The resulting drones will be needed to mate with young queens to be reared later. They take twenty-four days to become adult drones and are probably capable of service about a fortnight later. They are usually first seen in flight towards the end of April. The laying of the drone eggs in early spring (continued and increased through spring and early summer) is the first act of preparation for the swarm which is to come, and for the fertilisation of the young queen which will succeed the mother queen.

May is the month of blossoms. Nectar and pollen come freely into the hive, the population overcrowds the combs, the queen searches for empty cells in which to lay, and the nurse-bees, now very numerous, find themselves, actually or potentially, in possession of more secreted brood food than is needed. The general prosperity, congestion, and rising temperature combine to make the wax-building bees realise that the swarming time is at hand. They prepare for it by constructing a number of cup-shaped queen-cell bases on the combs in positions where there will be sufficient space for their later completion. This is the second act of the bees in preparation for the future swarm.

The third and decisive one is performed by the queen herself. She may pass and ignore the inviting cell-cups for some days, but suddenly, on a day that is warmer than usual, she will place an egg in each of them.

Having thus liberally provided for a successor, she "eases-up" in her laying for several days for she needs to be light enough to fly far and to have a reserve of energy for the great task awaiting her—that of departing with the swarm to found a new home, and to provide this with considerable quantities of brood before the advent of winter.

Meanwhile, what has happened to the eggs laid in the queen-cell cups? There may be any number of them

PLATE I

[Photo: L.E.S.

A Natural Swarm

from half a dozen to two dozen or more. Early on the fourth day from being laid they hatch, yielding tiny larvae which, resting on the bases of their cells, quickly receive from the nurse-bees small quantities of a rich secreted food commonly known as "royal jelly". They grow rapidly and receive increasing quantities of this royal food until they are five to five and a half days old. When sealed in their cells they have a reserve of food on which they feed during the pre-pupal stage (p. 43). In these important particulars they differ from the larvae destined to produce worker bees, for these receive a restricted and less rich food from the second larval day and do not feed during the pre-pupal stage (p. 72). The queen-cells are lengthened before being sealed until they are large enough to contain the full-grown larvae with abundant reserves of food, and later, the adult insects.

When the larva is five days old, i.e., on the ninth day from the laying of the egg, the open end of the queen-cell is sealed by the bees. The larva within, passing through the pre-pupal and pupal stages, during which many wonderful physical changes take place, becomes the imago or fully grown queen-bee, ready to emerge from her cell on the sixteenth day after the laying of the egg.

These dates—the ninth day and the sixteenth day—are of great significance in the raising of queens.

It is on the ninth day (sometimes the tenth) that the mother queen may be expected to depart with the swarm. On the sixteenth day the earliest of the young queens will emerge from its cell. Then one of two things may happen. The first princess to appear may search the hive for her sisters, whom she instinctively regards as rivals, and sting them to death whilst in their cells. Should others have emerged there will be mortal combats until only one survives.

Alternatively, if the stock is still reasonably strong in flying bees, one or more of the young queens may depart with small swarms. Finally only one young queen will remain in the parent hive. When she is about a week old she will fly to become mated. A few days later she begins to lay and becomes the new mother of the stock.

This is but a mere outline of what normally happens at swarming time. It is unnecessary at this point to amplify it. It must be realised that it is subject to variations on account of changeable weather and accidents. It serves however to explain nature's principal way of raising young queens, and at the end of this chapter we will consider what conclusions useful to the queen-rearer we may draw from it.

Let us now return to the original queen who has departed with the prime swarm. She finds a new home,—perhaps in the hollow of a tree or in a roof. In this the bees will build new combs and the queen will furnish them with brood. If the home is spacious the brood and stored honey will not fill it and the bees settle down and prepare for winter. If the space is limited, however, congestion and excess of brood food may recur, and the queen and bees may swarm again during the same summer. In such a case the queen is called upon to provide for three homes in one season. This may prove exhausting as her laying powers are limited. In either case we may assume that she and her bees go safely through another winter, and that during the following summer she passes through her second full season, the swarming taking place as before.

If she is of vigorous stock, and well bred and mated, she may pass through a third and even a fourth full season, but she is now becoming old. Her store of spermatozoa, derived from her drone-mate on her wedding-day, has greatly diminished, for she has expended several of them to fertilise each worker and queen-producing egg she has laid. (They have not been used for the drone-producing eggs.) And so she gradually lays fewer worker eggs and an increasing proportion of drone eggs which do not need fertilisation.

When this state of affairs becomes pronounced there appears to be a communal consciousness that the queen is failing and that the life of the stock is endangered. One queen-cell,—sometimes more than one—is constructed, and in it the old queen, perhaps instinctively conscious that she is sealing her own doom, lays the egg destined to produce her successor. The larva which emerges from this egg receives the most favoured attention of the nurse-bees and is liberally fed with royal jelly. As appropriate nutrition throughout growth is the most important factor in queen-rearing, this young queen is bred in ideal conditions and is likely to be of superior quality. Indeed, many beekeepers believe that supersedure queens, as they are called, are the best. It may well be that they are the best bred, but their usefulness will partially depend on other factors, e.g., parentage, and marriage.

When, in due course, the young queen emerges from her cell she finds the old mother queen parading the combs and laying a few eggs. Her natural impulse is to seek out and destroy her mother, but the bees often prevent this for a time, and it is not uncommon to find that the old queen has been preserved until the young one has begun to lay. Soon after this the mother disappears, either because food is withheld from her by the bees, or as the result of a sting inflicted by her daughter.

Let us consider the general effect of supersedure on a particular stock. When the bees decide to raise a new queen the old one is failing and is not providing enough brood for the continued existence of the stock. At least twenty-five days must pass before the young queen will be mated and ready to lay, and another twenty-one days must elapse before her first young bees are produced. For about seven weeks therefore the stock is diminishing in strength owing to lack of brood and at the end of this time it is in no condition to accumulate surplus honey.

Left to nature bees often supersede their queens at inconvenient times. Should this occur in March or April, or in September or October, there may be no drones, and the young queen failing to mate, the stock dies out. If in May or June, the stock may become too weak to take full advantage of the honey season. If in July or August, the stock may have passed through the honey season without great diminution, and favourable temperatures and the presence of drones will be conducive to the mating of the queen. During this part of the season the temporary lack of brood is an advantage rather than otherwise, and the stock, with ample time to raise brood before winter, is in a comparatively favourable state to bear the strain imposed upon it by supersedure.

As in all other cases of queen-rearing, there is the risk that the young queen may fail to become mated or that she may be lost on a mating flight. When this happens the stock perishes within the year unless it is provided with another queen by the beekeeper.

Queens suffering from disease, or from physical defects such as the loss of a leg or wing, queens whose wings have been clipped by the beekeeper, or those hastily reared and inadequately nourished in the larval stage, are often superseded by the bees before they become old, i.e., after one full season of service.

We may now consider the conditions under which queens are naturally reared, whether under the swarming impulse or in the course of supersedure.

Swarming normally takes place at the time of year when nectar, pollen, and water are coming plentifully into the hive. There is then a large population of bees of all ages, including the collectors of water, nectar, and pollen, the wax-builders, and a multitude of young nurse-bees of one to two weeks of age, best adapted to the secretion of larval and royal food. A great abundance of this food is available partly because there is some retardation of breeding during several days prior to swarming. Drones are numerous and on the wing. The teeming active population ensures the maintenance of the optimum hive temperature (93° F. to 95° F.) and the fanning bees at the entrance effect a constant changing of the air within the hive and so provide liberal supplies of oxygen without which breeding cannot be carried on. The most important point to notice is that the young queens are nursed and reared *as such* from the time when they hatch from eggs. The changes in their food (Chapter IV) which cause them to be differentiated from workers are made at the right times—which is not always the case in careless artificial queen-rearing.

In cases of supersedure also the young queens are reared as such from the hatching of the eggs. There is an abundance of larval food because few queen cells are raised and there is very little ordinary brood in need of food. The conditions of food supply and temperature may not, of course, be ideal when supersedure takes place too early or too late in the season.

From nature then we gather that the following are essential conditions for the rearing of good queens:—

(1) Bees of all ages should be present.
(2) Nectar (or honey, or sugar syrup), pollen, and water must be in plentiful supply.
(3) The population must be sufficiently dense to maintain a good hive temperature.
(4) Ventilation must provide for a large supply of oxygen—necessary in all brood rearing.
(5) The amount of young (unsealed) brood in the hive should be as limited as possible, this to ensure an abundance of available larval food for the young queens.

(6) The young queens should be reared as such from the day on which they hatch from eggs.

Many experienced beekeepers are convinced that other things being equal, the best possible queens are bred under the swarming impulse, for these are reared when the physical circumstances induce the strongest urge to natural increase, when prosperity reigns, and when the nurse-bees are not hurried in their duties. The conditions under which supersedure takes place may or may not be as good.

Why then do we not rely on natural swarming for the replenishment of our queens? Amongst good reasons are the following:—

(1) Modern beekeepers cannot be bothered with swarms. They are often lost and in any case cause much trouble.
(2) It is believed that continued breeding from queens that have swarmed perpetuates the tendency to swarm, but the view, held by some, that it increases this tendency, is doubtful.
(3) The utilisation of natural swarm queen-cells is not so simple as is sometimes represented (Chapter VIII).
(4) Beekeepers are loth to break stocks into nuclei just at the time when the honey season is opening.
(5) Young swarm-queens are ready for introduction to stocks whilst the latter are in the midst of the honey-gathering and when it is most inconvenient to disturb them.
(6) Opportunities for selection are limited, for frequently the best stocks do not swarm.
(7) There is much uncertainty as to the time of year when swarming becomes prevalent. In some seasons it is early, in others late; in some years swarms are plentiful, in others they are few. Everything depends on the weather and

income from March onwards. Even when a stock has made all preparations for swarming a spell of cold weather, strong winds, continuous rain, or even of drought, which brings scarcity of food, may cause bees to break down their queen-cells and defer swarming until a more favourable time.

In the next chapter we shall discuss how we may cause bees to rear young queens under conditions which approximate to those of nature, and shall consider more precisely the nursing and nutrition of queens in the larval stage.

CHAPTER III

LIFE HISTORY OF A QUEEN

Every animal and plant organism is made up of minute unit structures called cells. Although a cell is a highly complex structure, we may for the sake of simplicity and for the purpose of this chapter, regard it as a tiny mass of the substance of life called protoplasm, surrounded by a membrane—the cell wall—and containing within it a concentration of specialised protoplasm—the nucleus—which controls the functions of the cell and also carries within it much smaller bodies—the chromosomes—which ensure the continuity from one generation to another of characteristics which are described as hereditary.

The lowest forms of animal and plant life are unicellular. Each starts as a single cell, split off from its predecessor, and so continues throughout its life. Of this the amoeboid protozoa found in the sediment of stagnant water are a familiar example.

In the higher multicellular forms of life vast accumulations of cells go to the formation of the tissues of various body structures and functional organs. Growth and the renewal of tissues are effected by cell-division, one cell dividing into two, two into four, and so on. The cells of the same organism vary in size and form, according to their functions.

Two forms of cell with which we are here concerned are the ovum or unfertilised egg which develops in an ovarian tube of a fertile queen, and the spermatozoon which originates in a spermatic tube of a testis of a drone.

The egg.

The egg, at the time of being laid, is merely a very large cell. It is enclosed in an outer shell, the "chorion", which is thin and tough, and within which an inner skin or lining, the vitelline membrane, develops. It is about 1.6 mm. in length and .317 mm. in breadth at its larger end, and is slightly curved, being convex on the ventral side and concave on the dorsal side (56, p. *4*).

Viewed under magnification it is a beautiful object, being pearly white and translucent, and sculptured with a raised polygonal pattern which presents the appearance of a net. This pattern is discontinued at the anterior (larger) end, where is situated a tiny invisible opening, or perhaps a number of openings, called the micropyle. Through this enter a small number of spermatozoa pumped from the spermatheca of the mother queen (p. 52) as the egg passes down one of the oviducts (Pl. V, OvD) and through the common oviduct on the way to the genital aperture.

At this stage the egg contains amongst other organised materials:—

(1) A nucleus, situated near the anterior end and the micropyle.
(2) A mass of yolk cells destined for the nutrition of the future embryo,

and is undergoing a ripening process (p. 127). Its nucleus is already dividing into two daughter nuclei, one large and one small, which in turn again divide to form four.

Thus from the primary egg-cell nucleus four new nuclei arise. Of these, three ultimately disappear, leaving the largest, which becomes the female "pronucleus". This is capable of fertilisation and subsequent development into an embryo (24, p. *96*).

The spermatozoon of the male bee (Pl. II) is a cell

PLATE II

A. Living sperm from a testis of a young drone.

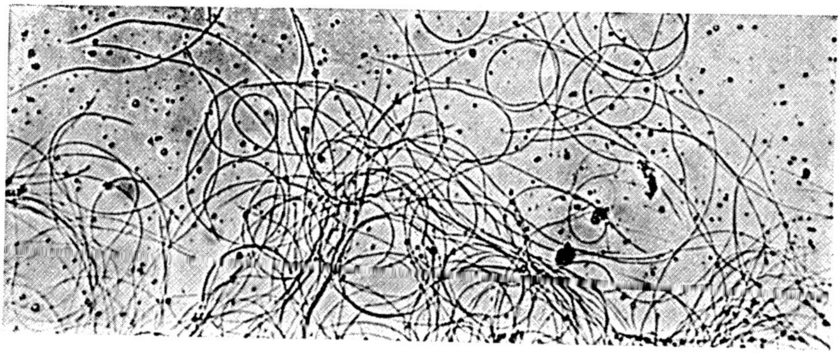

B. Living sperm from the *vesiculae seminales* of a drone.

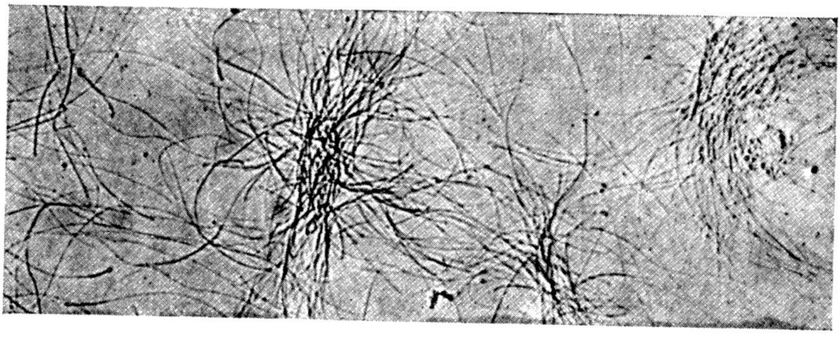

[*Photos: R. Bassindale*

C. Living sperm from the spermatheca of a queen.

SPERMATOZOA

of great length compared with its thickness. It comprises a needle-shaped head which consists of a little pointed piece and a larger back portion (the nucleus) and a long slightly flattened thread-like tail or lash which enables it to swim. Miss Betts (9, p. *43*), quoting Zander, gives its dimensions as 0.275 mm. long 0.0005 mm. thick. It is so small that the number of them comprised in the mass derived from a drone and stored in the spermatheca of a queen has been variously estimated as from four millions to two hundred millions. With reference to this point however Roberts (70, p. *225*) writes,—"Mackensen, (unpublished data), using a haemocytometer, has shown that the seminal vesicles of a mature drone contain about ten million sperm, whereas the maximum number found in the spermatheca of a queen was slightly over seven million."

A few spermatozoa, probably less than ten (9, p. *43*), only one of which finally functions, enter an egg as it passes down the oviduct, and it is their presence within the anterior end of the egg, immediately after this is laid, which is considered as evidence of the locality of the micropylar opening (or openings) which itself remains closed and invisible (56, p. *8*).

The head of the spermatozoon contains the male pro-nucleus, and the long tail contains the plasma, which in cells of other forms, usually surrounds the nucleus.

Within the egg the two pro-nuclei, that of the egg itself and that of one of the spermatozoa, coalesce, and form what is known as a segmentation-nucleus. This, by "cleavage" division gives rise to daughter cells and it is at this point that the life of the female bee begins (9, p. *24*). These cells themselves divide, and so on in rapid progression until the vast numbers of cells needed for the formation of the embryo and subsequently of the various organs and body structure of the new individual are available.

To attempt to describe the complicated processes involved in the formation of the embryo would be out of place here. For a full account of these the reader is referred to Nelson's "Embryology of the Honey Bee", or for a good summary of them, to Miss Betts' "Practical Bee Anatomy". The main point with which we are here concerned is that when at the end of 74 to 76 hours the tiny larva bursts its shell its internal organs, including its ovaries, already exist in rudimentary form. On its subsequent nutrition and the consequent development of its sex organs depends whether it will become a worker or a queen bee.

When first laid the egg is attached at its smaller end to the base of the cell by means of a minute quantity of adhesive substance secreted by the queen (56, p. *5*) which maintains it in a position perpendicular to the base of the cell. If this adhesive be broken the egg cannot be restored to the same relative position in another cell, and for this reason we cannot transfer detached eggs from one cell to another for the purpose of queen rearing, although the bees are believed by some to be able to do so (73, p. *55*). On the second day the egg becomes slightly inclined to the cell base and on the third day, a few hours before it hatches, it lies prone on the base of the cell. These successive positions enable us to determine the approximate age of the egg, which is important when we wish to select cells containing 3-day-old eggs for queen-rearing, e.g., in the Barbeau method (Chapter XVI).

Just before the egg is due to hatch, i.e., at the beginning of the 4th day, the nurse-bees deposit near it a tiny quantity of larval food (p. 34). If this is not provided, as is sometimes the case in times of scarcity, the egg, according to Root, does not hatch (73, p. *98*).

Leuenberger (46, p. *146*) however found that normal eggs, prevented by wire gauze from receiving brood-food, hatched at the proper time, but that the resulting larvae appeared to be dead. The same writer (46, p. *167*) investigating cases of queens producing so-called sterile eggs found:—

(1) That the drop of larval food was generally lacking.
(2) That sterility was not cured by transferring the eggs to normal stocks where they might receive larval food.

He surmised that probably the bees did not provide food for sterile eggs knowing that the embryos within them were already dead, and that their deaths might be due to some disease of the queen.

The larva.

About 76 hours after the laying of the egg the embryo bursts its shell and emerging as a tiny larva takes up a slightly curved position in its little bath of food. At this stage it is very small, being not quite 2 mm. in length and is hardly distinguishable from the food with which the bees immediately surround it. It is then too small and too delicate to be successfully removed to another cell. It eats the secreted food supplied by the nurse-bees, assumes the form of letter C and grows so rapidly that, according to Nelson and Sturtevant, cited by Phillips (64, p. *111*), who give a table of average weights of worker larvae on successive days, its original weight is increased 6½ times during the first 24 hours and 1,576 times during the 4½ to 5 days of its existence as an unsealed larva. This, in comparison with the growth of other animals, is phenomenal, and depends on frequent administrations of minute doses of food provided by the nurse-bees. This food varies, as will be shown later, in quantity and quality according to whether it is supplied to larvae intended to become queens or workers (Chapter IV). It is ingested by the growing larva through its mouth, and not, as some state, by absorption through its skin. As its stomach is not yet connected with its intestines (80, p. *299*) the whole of its food is absorbed and therefore at this stage there are no faeces.

Lineburg (73, p. *101*) gives an interesting description of the larva's circular movements which enable it continually to meet fresh food and to keep the whole well mixed. He computed that the total number of visits by nurse-bees to a single larva during its 5 days of unsealed existence is about 10,000. Even if we do not suppose that the number is much greater for the nutrition of a queen-larva the amount of secreted food needed is vastly greater, and we can therefore readily appreciate why we should never attempt to rear queens in weak stocks, especially those with few or no nurse-bees.

Whilst floating on its side in its bed of brood-food the larva breathes through the ten spiracles on the upper side of its body, the corresponding spiracles on the under side being immersed in the fluid food and therefore out of use. For this reason care must be taken in grafting (p. 241) to place a transferred larva lightly on the prepared royal jelly in a similar position to that which it previously occupied. If it is inadvertently immersed it cannot breathe and will die.

At the beginning of the third day the nurse-bees change the diet of the larvae which are to become workers by adding honey and undigested pollen to it. At the same time there is a marked change in the chemical composition of the secreted food (Chapter IV). The effect of these changes is to impede growth, and in particular the development of the ovaries, and the larvae are said to be weaned.

In the case of a larva destined to be a queen the rich diet of secreted food is continued, greatly increased, and modified in composition (p. 66), with the result that the ovaries develop without retardation and attain their full size—as of a virgin queen (Pl. III, B)—by the time the queen reaches the adult stage.

The ovaries of the adult worker however are atrophied and very small, appearing like two threads hardly visible to the naked eye (Pl. III, B). Each usually comprises about six ovarian tubes but the number is variable. Haydak has found as many as 59 in the two ovaries of a worker (27, p. *757*). There is no functional spermatheca (p. 52) and the insect is incapable of being effectively mated and of laying fertilised eggs.

Each of the queen's ovaries normally comprises from 160 to 180 ovarian tubes, which, when filled with developing eggs in the fertilised queen, are very large, and occupy a considerable portion of the abdominal cavity. In the young unmated queen the genital organs, including the spermatheca, are fully developed and adapted for sexual union.

From the beginning of the third larval day the quantity of brood-food, now called royal jelly, supplied to a queen larva is greatly increased by the bees. It is known that bees of all ages are able to produce this food—a secretion of the pharyngeal glands. These glands however are most active and productive in bees of six to fifteen days of age which for this reason are usually described as nurse-bees. Snodgrass (80, p. *171*) describes royal jelly as "A gummy paste of a milky white colour when fresh, though when removed from the cell it soon acquires a darker tone with a yellowish tint. Under the microscope it appears to be a homogeneous, very minutely granular substance. It is strongly acid, and very acrid and pungent to the taste".

Towards the end of the fifth larval day the queen-grub receives its last increment of food and is sealed up in its cell. It then has to spin its cocoon and subsequently to pupate, processes involving the consumption of food in the case of the queen-larva which

PLATE III

[Photos: R. Bessindale

OVARIES

A. Ovaries, paired ovarian ducts, spermatheca, spermathecal glands, and sting glands of a laying queen.

B. Comparison of generative systems of a laying queen, a virgin queen, and ordinary worker bees.

actually increases in weight after the cell is sealed. The worker-larva, on the contrary, spins its cocoon and pupates in a state of starvation, and therefore loses weight during the prepupal stage. If the queen-larva has received the full attention of sufficient nurse bees it goes into its prepupal stage with such a large reserve of food that some of it remains in the cell after the adult insect has vacated it.

If queens are reared in weak or hungry stocks the absence of this reserve of food is observable before the cells are sealed and may be considered a sign of deficient larval nutrition and consequent retardation or prevention of bodily growth and ovarian development. The resulting queens will be small and unprolific, if not useless.

When, in the prepupal stage, the larva has taken its last meal the stomach and intestine of the larva become connected, and its excreta are deposited at the base of its cell. Still a larva (prepupa) it straightens itself along the length of the cell and on the first day spins a silken cocoon, which in the case of the queen, lines the anterior portion of the cell and is supplemented by a thin membraneous lining of the sides but not base of the cell (9, p. 27). Enveloped in this covering, which now lines the cell, the larva rests for a day and then the greatest change of all, metamorphosis, takes place. The larval skin is shed—the fifth and last larval moult—and the grub appears as a pupa—replete with external limbs and internal organs, all white, delicate, and immature, and enclosed in a transparent membrane.

The pupa.

The pupal stage lasts, in the case of the queen, for five days. During this period numerous and profound changes take place especially in the internal organs, some of which are broken down and reconstructed to assume the form they take in the adult bee. For detailed reference to these changes the reader is referred to books on bee anatomy (80 and 9).

On the fourth day the pupa casts its skin and emerges from it as an imago or perfect insect. It is now almost ready to leave its cell. Anticipating this the bees then thin down the queen-cell wall in a circle just above the tough cell-end. This circle is darkish in colour and serves to indicate to a beekeeper that the queen is likely to emerge from the cell within a few hours.

The adult queen.

On the sixteenth day from the laying of the egg from which she has developed the queen bites through the thinned circle of the cell wall with her sharp mandibles. The cell capping falls away except for a small portion which retains the cap in the manner of a trap-door, and the young queen walks out on to the combs. The presence of the hinged-cap at the mouth of a vacated cell is a reliable indication that a young queen has emerged. Often however the cap falls away entirely. In rare cases it may close and be re-sealed by the bees over the empty cell.

Queen-cells.

The appearance of a queen-cell usually affords a clue as to the condition of its inmate. A normal cell, not built in haste, projects somewhat from the face of the comb, hangs downwards to the extent of about an inch, is tapering, and has a corrugated appearance, the corrugations near the base having the form of incipient worker cells. The basal portion of the interior of the cell, originally part of the comb, is usually filled with royal jelly, and the young queen develops and matures in the part external to the comb.

A cell made in haste over a larva of the second or third day, as when a stock is suddenly made queenless, is small and often scarcely projects from the comb, and the queen within it is worthless.

A cell which has been greatly lengthened may contain a larva which has not developed normally and may be dead.

A cell with greatly thickened walls probably contains a dead queen, the bees having added wax to the walls whilst waiting overtime for the queen's emergence.

A smooth-walled cell is suspect even if of normal size. When the bees construct queen-cells over drone larvae, as they occasionally do in normal circumstances, and frequently when laying workers are present, the cell walls are usually smooth.

Queen-cells contiguous to sealed drone brood must be suspected, although if their walls are well corrugated they are likely to contain good queens.

Occasionally a worker bee is imprisoned in a queen-cell from which a queen has recently emerged. For some reason the hinged cap falls, or is pushed back into its place, and the bees, in error, reseal it.

It is well therefore, when selecting a particular cell from amongst a number, to choose it before it is sealed and when the queen larva can be seen.

Herrod-Hempsall (33, p. *85*) describes how the inmate of a cell may be inspected by cutting away a slice of the wall near the base with a sharp warm pen-knife. The slender larva or pupa of a queen is readily distinguished from the more bulky one of a drone. After inspection the excised part of the wall is replaced and its edges smoothed down with the warmed knife. The bees then complete the repair of the cell and its inmate is unharmed.

At the time of natural swarming bees sometimes prevent young queens from leaving their cells on the sixteenth day, thus preserving them for issuing with second and subsequent swarms. A queen so imprisoned, unable to reach the residue of royal jelly at the base of its cell, would die of starvation but for the nurse-bees which pierce a small hole in the cell-wall just above the capping and through it feed

PLATE IV

[Photo: H. Teal

NATURAL SWARM QUEEN-CELLS

her with brood-food. The imprisonment may last for three or four days.

Rivalry of queens.

A queen will not tolerate a rival in the same hive. On rare occasions two queens have been known to live and lay in different regions of a hive, and in cases of supersedure the young queen sometimes begins to lay before her mother ceases work and is disposed of. In such cases the queens are seen to steer clear of each other, and sometimes they are kept apart by the bees. If they meet they grapple with each other, each trying to get into a favourable position for administering a death-blow with her sting. Suddenly one seems to relax her efforts and is immediately stung to death by her rival. Usually the younger and more vigorous is the victor.

When a young queen emerges from her cell she first helps herself to food and then searches for her sisters, whether emerged or still in their cells. With the former she engages in deadly combat. Finding the latter she bites a small hole through the wall of each cell. This incites the workers to break down more of the cell wall and to remove the immature queen, often piece by piece. It is commonly believed that the first queen to emerge stings her younger rivals to death whilst they are still in their cells, but according to Root (73, p. *544*) there is some doubt about this. Pease (60, p. *10*) has proved that if she is prevented by excluder from reaching them the bees tear down the cells and destroy the inmates.

If a number of young queens are to be preserved for after-swarming the sororicidal attacks of the first-emerged queen are prevented by the bees for a day or two, or even longer. Finally of those remaining all except one will be destroyed,—perhaps an example of "the survival of the fittest".

It is for this reason that a queen-breeder takes the precaution to transfer his queen-cells to nuclei or nursery cages on the tenth day when the queens are not expected to emerge until the twelfth day. Should one of the queens emerge a day earlier than expected she would quickly destroy all the remaining queen-cells.

We may here remark that the short curved sting of a queen is reserved for use against a rival queen. Only very rarely has she been known to sting a human being.

Comparison of pre-adult stages of queen and worker.

	Queen.	*Worker.*
Incubation of egg	3 days	3 days
Egg hatches	Beginning of 4th day	Beginning of 4th day
Larval stage	5 days	5 days
Cell sealed	9th day	9th day
Pre-pupal stage	2 days	3 days
Pupal stage	5 days	9 days
Change to imago	15th day	20th day
Imago emerges	16th day	21st day

The Virgin Queen.

Both before and after emerging from its cell the virgin queen is able to call to her rivals in other queen-cells or to the mother queen, who may reply, by means of a pleasantly shrill sound resembling that of a child's toy trumpet. This sound is frequently heard from outside the hive prior to the issue of a swarm which has been delayed by bad weather. This "piping" as it is called, is regarded by some as in the nature of defiance or challenge, but it may also be emitted by a queen when apparently in a state of terror.

The newly emerged queen is slightly moist, downy, full-sized, and at once active in movement. She soon helps herself to a good feed of honey and then roams over the combs, little noticed by the bees which do not make way for her or attempt to feed her as in the case of a fertile queen. She quickly dries, shrinks a little in size and becomes extremely active. When disturbed she shows nervousness and tries to hide amongst the bees, and if frightened by successive and unskilled attempts to catch her she takes flight (p. 323). She feeds herself on honey and pollen until she is mated and ready to lay, after which time she is fed by the nurse-bees. Whether during her virginal period she receives any supplementary food from the nurses, or helps herself to it, is very doubtful, although it has been suggested that she does (p. 302). At all events she is quite well able to support herself for some days on honey and pollen, or, when caged, on queen-cage candy with a small admixture of pollen.

On or about the fifth day after emerging, weather permitting, she takes her first practice flight, which is succeeded by others. Meanwhile her generative organs have been maturing and a day or two later she attains a state of aphrodisia which impels her to take her first mating flight. It is possible that this state waxes in intensity at intervals and that during the intervening periods the queen does not seek to be mated. If this is so it may account for the observed fact that fine warm days do not necessarily ensure mating.

During her virginity the queen is able to feed at will provided the hive contains stores of both honey and pollen. It is a matter of common observation that she is much more likely to become mated from a strong, well-provisioned stock than from a weak and hungry one. Both temperature and food appear to have important bearings on her physical economy. Most writers who describe nursery cages (p. 300) recommend a provision of sugar candy alone for the imprisoned virgin queen, —an imperfect food at a critical stage of her life (p. 301).

Fecundation.

Bad weather may delay, or may cause to be unsuccessful, the early mating flights of a virgin queen, but within the first month of her life she continues to take them until she is mated. Some account of these flights and the actual mating is given in Chapter V.

It has long been the general view, not shared however by a few writers, that a queen mates only once in her lifetime, and that at this single mating she receives and stores in her spermatheca sufficient spermatozoa for the fertilisation of all worker-producing eggs she will lay in the course of her life of three or four years. This is true in a considerable proportion of cases but it is now well established that many queens mate more than once. It has been conjectured that the desire to mate a second time may be due to some temporary non-functioning of the generative mechanism of the queen or to the fact that the spermatheca is not filled at the first mating, suppositions which to some extent appear to be supported by the researches referred to below.

It seems probable however that a queen bee has a period of maximum receptivity and that whilst the state of aphrodisia is intense, multiple mating, although not necessary for complete fertilisation, is to be expected.

In the June number of "Gleanings in Bee Culture," 1944, William C. Roberts, (70, pp. *255–303*) of the U.S. Department of Agriculture, who had collaborated with Mackensen in some of his work, describes a series of extremely instructive experiments relating to the multiple mating of queen bees. As only limited reference can be made to them here the reader is recommended to study the whole article.

In the course of these experiments marked Italian virgin queens in nucleus hives were prevented from flying—except under observation—by excluder guards placed over the entrances. The hives were watched daily from 12.30 to 5 p.m. during which period queens and drones usually fly. When a queen appeared at an entrance the guard was removed and the times of leaving and returning, and evidence of mating (p. 92) were recorded. Subsequently the colouring of the progeny was noted as indicating male parentage.

Amongst the facts confirmed or established were the following:—

(1) Few queens appeared at the hive entrance before the age of six days.
(2) Of 110 marked queens allowed to fly during observational periods 55 mated twice.
(3) The second mating generally occurred on the day after the first, but one queen mated twice on the same day.
(4) 14 other single-mated queens made additional flights after mating.
(5) The average duration of mating flights, as distinguished from shorter non-mating flights, varied from 19.3 minutes in April to 11.9 minutes in June (when drones were more numerous).
(6) Of 35 yellow queens allowed free flight at mating stations provided with yellow drones, eight mated only with yellow drones, one mated only with black drones, while 26 mated with both yellow drones and stray black drones.
(7) Under natural conditions the proportion of double matings was considerably greater than under the controlled conditions.
(8) The double-mated queens appeared to be more productive.
(9) Distances of 1¾ miles and 2 miles were not sufficient to isolate the mating apiaries from others containing black bees.

Roberts concluded that multiple matings by queens appear to be more frequent than single matings.

For the queen-breeder these experiences are significant in that they point to the necessity of effective isolation and an abundance of selected drones in a queen-mating apiary. They suggest too, that single matings may sometimes be imperfect, which may account for observed limitation of the duration of fecundity in some queens, and their consequent supersedure.

Plural mating accounts for the observed production from one queen of different kinds of bees, one kind sometimes succeeding the other soon after brood rearing has commenced. It also affords an explanation of an apparent deviation from parthenogenetic inheritance noticed by Doolittle (p. 150).

In a recent report of work in the Bee Department at Rothamstead (September 1945), Dr. Butler states that "considerable progress has been made in studies of honeybee biometrics designed to discover simple and rapid methods of determining and describing the race and strain of any group of bees taken from any one colony. It has already been found practicable to use biometrical methods to determine whether any given queen has mated with any desired strain of drone, or even drones, even though such matings are not under control". This, he adds, should facilitate line-breeding and the production of strains of bees suitable to particular environments and systems of management.

Duration of periods of receptivity.

When receptive a virgin queen will fly even if the weather is not sufficiently good to incite the flight of drones. Unless therefore her period of receptivity coincides with good weather she may fail to mate and as she appears to lose sexual desire at the age of three to four weeks she will, if she lives, become a drone-breeder.

A queen may mate as early as the sixth day after she emerges from her cell. Observations of 60 matings at Baton Rouge, La., U.S.A., in 1937 and 1938 (73, p. *205*) showed that the greatest numbers, viz., 16 and 16, were effected on the 8th and 9th days respectively.

Egg laying began from one to eight days after mating, the greatest numbers of cases being,—after two days 11, after three days 21, and after four days 13.

In 1787 Huber (36, pp. *32–33*) in his experiments prevented fecundation for varying periods and stated—"I can always prevent queens from laying the eggs of workers by retarding their fecundation until the twenty-second or twenty-third day of their existence . . . she then begins to lay the eggs of males and produces no other kind during her whole life". Not being aware of parthenogenesis (p. 124), which was discovered later, Huber was greatly puzzled by this phenomenon, and concluded by writing "It is an abyss wherein I am lost".

It is popularly supposed that a virgin queen cannot become fertilised after she is 28 days old. Of this Morison (52, p. *72*) referring to his own observations, gives the following explanation:—

> The spermatheca of a young virgin queen is filled with a fluid in appearance something like the white of an egg. After mating the mass of spermatozoa derived from the drone gradually enters the spermatheca, displacing an equal mass of the fluid. If however the virgin queen fails to mate this fluid changes to a solid state thus making the entry of spermatozoa impossible. The solidified substance can be extracted like a small glass bead. The change takes place gradually the rate probably varying in different queens, so that the age at which a queen becomes incapable of fertilisation is not rigidly fixed.

In his experiments on instrumental insemination Watson (92, p. *44*) succeeded in inseminating, partially or fairly copiously, queens of 35 to 37 days of age.

We may presume that the physical change noted by Morison would be contemporaneous with the cessation of sexual urge and that the queen would not subsequently seek to be mated.

It is a matter of common observation that queens mated towards the end of their first month of age are unprolific and are liable to early supersedure.

Calm warm weather, or warm showery days with sunny intervals are conducive to mating. Cold weather or high winds are unfavourable. Some seasons are distinctly more favourable for queen-mating than others.

Huber observed that his queens commenced laying worker eggs 46 hours after being mated. This is now recognised as usual, but in the case of queens reared in the autumn, cold weather and lack of income sometimes delay the laying of the first eggs for some time after mating has been accomplished. A queen should not be condemned as unfertile until her stock has been provided with warm syrup which quickly induces her to lay if she has been mated.

Occasionally a mated queen lays a proportion of drone-producing eggs at first. She should not be condemned until it is seen whether, within a few days, she produces worker brood in the normal way. In other cases the appearance of a small proportion of drone brood amongst the earliest worker brood may be attributable to the last efforts of laying workers.

Controlled fecundation of queens.

The impossibility of selecting, under natural conditions, the drone which shall mate with a particular queen has always been and still remains the chief obstacle to general progress in improvement of bees.

Since the days when Reaumur, about 1740, placed a virgin queen with several drones in a glass jar, hoping to observe the act of union, numerous experimenters have, in diverse ways, endeavoured to secure fecundation under artificial conditions.

An interesting account of some of these attempts is

given by Lloyd R. Watson (92, pp. *9–16*). They include the following:—

(1) Allowing virgin queens and drones to fly together in enclosed spaces, varying in size from the interior of a glass vessel to that of a huge greenhouse.
(2) Tethering a virgin and a selected drone by fine silken threads which, while allowing them to meet, limited their range of flight.
(3) Expulsion of the seminal fluid of immature drones on to queens in the pupal stage.
(4) Dropping expelled seminal fluid into the genital aperture of a virgin queen.
(5) Smearing the micropylar ends of newly laid drone eggs with sperm.
(6) Forced copulation effected by holding queen and drone in suitable contact whilst pressure is applied to cause eversion of the drone organ.
(7) Direct injection of spermatic fluid into the vulva of the queen by sufficiently refined instruments and technique.

Although occasional successes of some of these experiments have been claimed, the great majority of them have proved to be total failures and biologists have not been able to confirm the reported successes.

In 1926 Watson (92, Chapter V et seq.) after long and patient work succeeded in devising a refined technique which enabled him to effect the artificial insemination of queens. Since that time his apparatus and procedure have been simplified and improved, and his method is now used by skilled biologists in connection with stock improvement and studies in heredity.

Briefly Watson's method was as follows:—

The virgin queen to be inseminated is placed on her back in a cradle, hollowed out of a convenient wooden block (operating table), of a shape to fit her body comfortably, and is held down by means of silk loops passed over her body and around the block. The block is so shaped and adapted to the stage of a low-power microscope as to give an upward slope to the queen's abdomen the tip of which is brought into the field of the microscope where it is illuminated by a strong beam of reflected light.

A micro-syringe the plunger of which is controlled by means of a fine screw, mounted in a syringe-holder by which it is mechanically adjusted to the line of the queen's abdomen, is charged with mucus and semen taken separately and successively from the bulb of the penis of a selected drone.

The abdominal tips of the queen's abdomen are held apart by means of tweezers and the nozzle of the syringe slowly advanced until it just disappears into the vulva of the queen. The semen, followed by mucus, is then slowly expelled from the syringe by advancing the plunger.

The mucus quickly coagulates on exposure to air and forms a plug which prevents the escape of semen after injection.

In this brief account of Watson's technique several refinements found to be necessary are not mentioned and for these the reader is referred to Watson's book on the subject. It may be sufficient however to indicate that the whole procedure is extremely delicate and suitable only for a highly-skilled operator.

Watson records that in an early series of experiments he succeeded in inseminating at least 57 per cent of the queens operated upon, the degree of insemination varying from "slight" to "fairly copious" as shown by subsequent dissection. Some of the queens were introduced to nuclei where they produced worker brood.

Artificial insemination of queen bees is not as complete as in nature and it differs from that of most other animals in that the amount of sperm injected and stored is of prime importance. Its chief value lies in the facilities it affords for the perpetuation of desirable characteristics by in-breeding or crossing, and in the provision of controls for experiments in other directions, —e.g., those referred to on pp. 45–46. It seems unlikely that it will ever be used by the amateur or commercial beekeeper who must continue to rely on what degree of isolation is attainable to secure desired matings of his queens.

For an interesting account of improvements in the technique of the artificial insemination of queens the reader is referred to a recent article on the subject by H. H. Laidlaw, Jun. (44, pp. 429–465).

In a recent communication to the writer Dr. Butler states that the technique of instrumental insemination has now been so much improved in America that it is being used in the production of queens for honey production trials, the work being carried out by Dr. Mackensen.

Fertility.

After mating, the ovaries develop rapidly until they occupy a large part of the abdominal space and appear as illustrated in Pl. V. Each ovary comprises about 180 ovarian tubes (Pl. V, ov) and each of these contains a number of eggs in different stages of development. Each egg originates as a tiny cell in the fine end of the ovarian tube and is accompanied by a small mass of nurse cells by which it is nourished and which it finally absorbs. It grows in size as it passes down the tube until it emerges a full-sized ovum or egg into one of the ovarian ducts (Pl. V, OvD). These unite to form the common oviduct in which the egg passes the opening of a short duct which leads from the spermatheca (Pl. V, Spm).

The spermatheca is a small round sac, of about 1 mm. diameter, in which is stored the mass of spermatozoa derived from the drone in mating. It is situate above the common oviduct and is conspicuous on account of its glistening surface due to the close

PLATE V

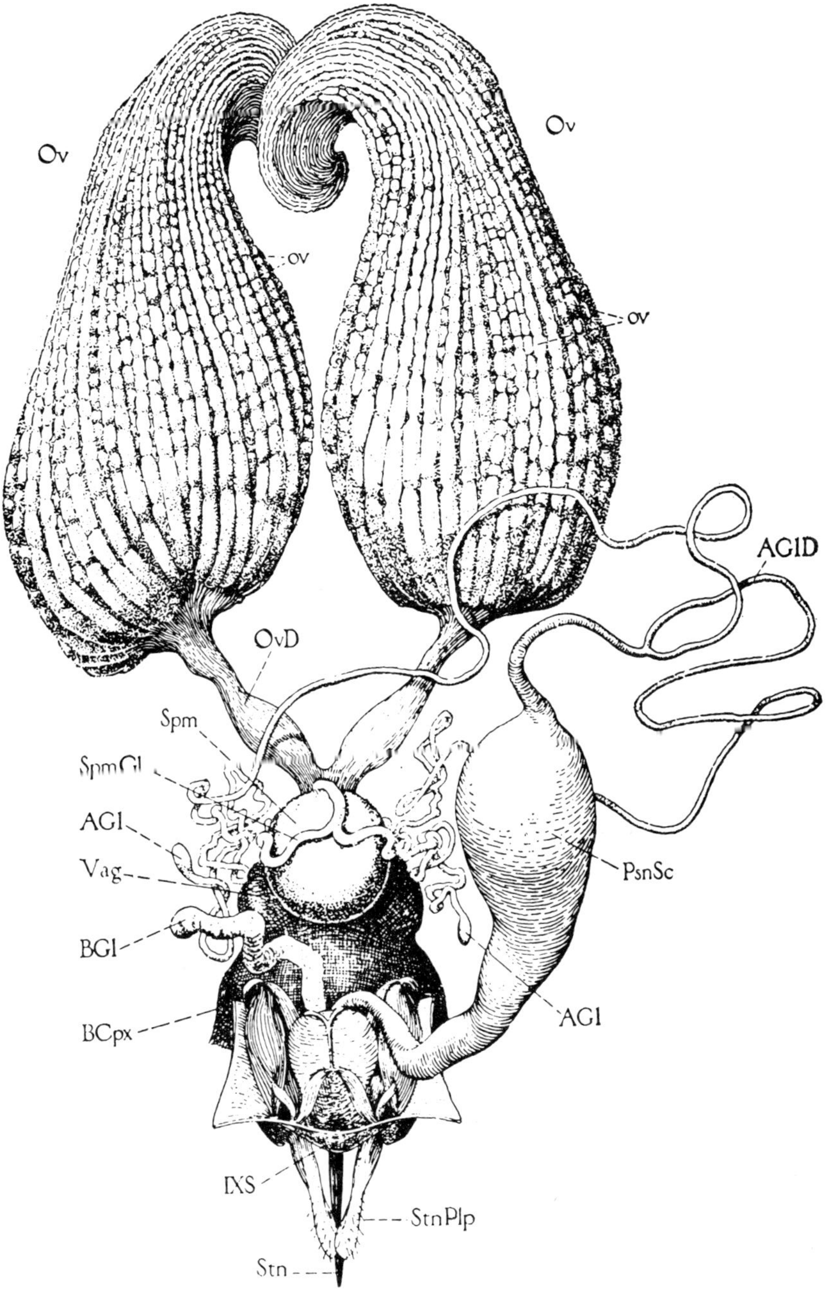

[*By courtesy of the McGraw-Hill Book Company, New York*

THE REPRODUCTIVE ORGANS, STING, AND POISON GLANDS OF THE QUEEN, DORSAL VIEW. (After Snodgrass.)

net-work of tracheae which covers it. The short duct leading from it to the oviduct serves as a passage for the entry of the spermatozoa after the mating of the queen. It also regulates their disbursement for the fertilisation of passing eggs by means of a muscular pump, first described by Breslau (80, p. *260*). The spermathecae of a virgin queen and of a fertile queen, together with their glands, are shown in the photograph in Pl. III, B.

If the passing egg is destined to produce a female bee it must be fertilised and this is effected by the pumping of a small group of spermatozoa on to it as it passes, some of which enter it at the micropyle (p. 30). If the egg is to produce a drone this fertilisation does not take place.

For a fuller description of these processes the reader is referred to Snodgrass' "Anatomy and Physiology of the Honey Bee", Chapters XI and XII.

A glance at the illustration (Pl. V) makes it easy to realise that a fertile queen is able, at her peak period of laying to produce an average of about 1,500 eggs per day for many days, with a maximum of about 2,000 in one day (72, pp. *109–110*). She lays throughout the 24 hours, taking short periods of rest during which she is fed with secreted food by the nurse bees. In the winter she receives less food, her ovaries contain few eggs and diminish in size, and laying ceases until the early days of the new year. After two full seasons her stock of spermatozoa is greatly reduced and she lays fewer fertilised eggs. Growing older she begins to lay an increasing proportion of unfertilised (drone producing) eggs and her stock becomes weak. She is then superseded (p. 24).

Queens are considered to vary in fertility partly through restricted or defective nutrition in their pre-adult stages, and partly on account of the completeness or otherwise of their fecundation. Some (e.g., those of certain Mediterranean races) lay abundantly and

PLATE VI

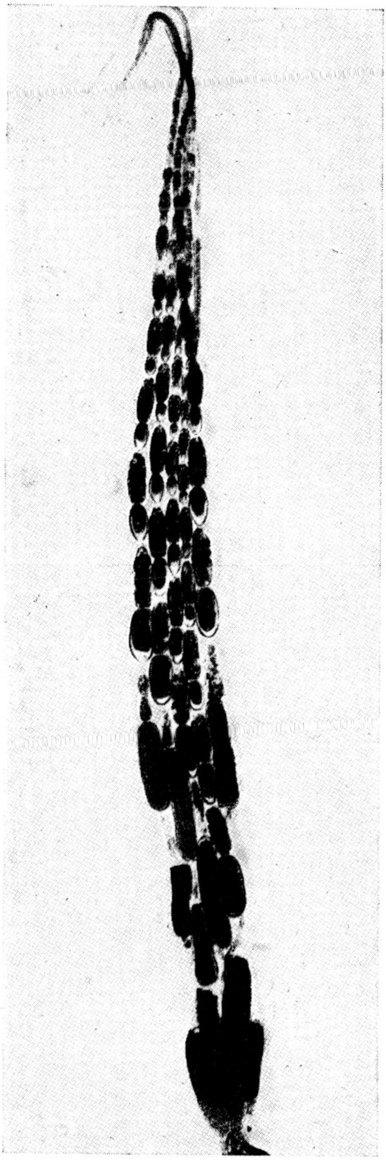

[*Photo: R. Bassindale*

FIVE OVARIAN TUBULES (OVARIOLES) FROM AN OVARY OF A FERTILE QUEEN SHOWING DEVELOPMENT OF THE EGGS, WITH SOME ASSOCIATED NUTRITION CELLS. THE TWO OVARIES COMPRISE ABOUT 350 OF THESE TUBULES throughout a great part of the year, sometimes producing enormous stocks. The queens of the dark varieties of cool climates are more restrained in laying and their stocks work with smaller populations, but nature seems to provide some compensation for this in that their bees have longer lives (1, p. *84*).

Poorly-bred or imperfectly mated queens are often not prolific at any time and fail early in life; consequently their stocks are unprofitable.

A queen reaches the peak period of her laying in a particular year just before swarming time, the rate varying approximately as the temperature and the amount of food given to her by the workers. In times of low temperature and scarcity this food may be largely withheld and laying is then correspondingly reduced or may cease altogether.

To distinguish young and old queens.

The characteristic lively and nervous movements of a virgin queen are described on p. 44. It is often desired to distinguish between a young and an old fertile queen. This is not always easy, especially in the autumn when a fertile queen may have ceased laying.

A young queen keeps her wings neatly folded over her body and the tip of her abdomen clear of the combs when walking. She moves quickly when disturbed, and may take flight like a virgin if agitated by ineffectual attempts to catch her. Her body, and in particular her abdomen, has unimpaired pubescence which although not markedly visible, gradually wears off as she advances in age.

An aged queen is large, slow in movement, and sometimes, when in full lay, is unable to keep the tip of her abdomen clear of the combs. Her wings tend to spread apart and are often frayed at the ends. She is little perturbed by sudden disturbance and can easily be caught. The constant insertion of her abdomen into the cells in the act of laying has worn away the hair fringes of the segments with the result that her body appears naked and shining. When she has ceased laying however, as in the autumn, she becomes smaller in size, neater in appearance, and more active in movement.

Whilst it is not difficult to distinguish between a very young and an aged queen, it is not easy to recognise a middle-aged queen by appearance only. For this reason a system of marking is desirable.

Marking queens.

A spot of brightly coloured rapidly drying enamel paint, applied to the dorsal surface of the thorax with the point of a fine camel hair brush will not only facilitate the search for a queen but will serve to indicate her age if different colours are used for different years.

The Ekhardt system of marking is better, however. It involves the use of a small punch and plunger and consists in the fixing, by means of the latter, a brightly coloured, gummed, and numbered disc of metallic foil to the thorax of the queen. This, if properly affixed, she will carry through life. The outfit, as sold, comprises a punch with plunger, sheets of metallic foil variously coloured and numbered, gum, and directions.

The operation is easy. It is not necessary to repeat the directions here but the writer offers a suggestion for the holding of the queen whilst the disc is being fixed. If held by one wing only between the thumb and first finger, she will turn until all her feet are firmly planted on the finger, and will remain perfectly still whilst the operation is performed.

It is usually unnecessary to mark virgin queens as some of them may fail to mate. Manley (50, p. *71*) suggests that the marking with discs may have an adverse effect on introduction and recommends that a fertile queen should be marked after she has been introduced and has established a brood-nest.

As the holding of the queen demands much care the beginner should first acquire some skill by marking a few drones.

The queen in disease.

The queen, in common with drones and workers is subject to the diseases of adult bees, but is apparently unaffected by diseases peculiar to the brood.

Acarine disease.

In Acarine disease the queen is often amongst the last of the family to die although herself heavily infected (68, p. *14*).

Nosema disease.

In Nosema disease she may not only be heavily infected, but in this state will become a centre from which the disease spreads on account of her infected excrement having to be removed by the bees. Frequently she dies at a comparatively early stage of the development of the disease in her stock (69, p. *35*).

Fyg, cited by Miss Betts (14, p. *5*), has shown that Nosema infection of a queen puts a large number of the ovarian tubules out of action. He concludes that if the queen of a Nosema infected stock maintains her normal rate of laying it may be assumed that she herself is not infected.

Bee Paralysis.

At least one of the states collectively known as "Bee Paralysis" appears to be caused by some defect in the queen, for re-queening from a healthy stock effects a permanent cure.

Addled brood.

Similarly the condition known as "Addled Brood" is attributed to some defect in the queen (84, p. *2*) and is alleviated or cured by the substitution of a queen of healthy antecedents for the queen of the affected stock.

European Foul Brood.

In European Foul Brood a period of queenlessness, followed by re-queening from a healthy stock, is favourable to recovery because it gives the bees a chance to remove the dead diseased brood, but a cure of the disease does not necessarily follow.

A queen is not known to be infected in either European or American Foul Brood. Should she be specially valuable she may be saved when her stock is destroyed. If given an escort of bees from a healthy stock and kept in a clean cage for a day she may be introduced to another stock without danger of communicating the disease. Care must be taken of course that honey used in the making of queen-cage candy (p. 300) has come from healthy bees.

Melanosis.

Queens both young and old may be affected with the disease known as Melanosis, (11, p. 59), sometimes called "black egg" disease. The tissues of the ovaries, and at times of other organs, degenerate and turn black probably as a result of invasion by a yeast-like fungus which is present. Queens affected either do not begin to lay, or cease laying altogether after having laid normally.

Generally therefore we may say that queens should not be saved from stocks affected with Acarine Disease, Nosema Disease, Paralysis, Melanosis, Addled Brood, and that only in exceptional cases should they be preserved from stocks affected by Foul Brood.

Queen bees are sometimes affected by certain minor diseases, deformities, or other abnormalities, including those resulting from injuries. For an interesting summary of these the reader is referred to The Apis Club's book on "The Diseases of Bees" (85, p. *28*).

The Braula coeca.

The Braula coeca which sometimes infests individual bees of a stock seems to have a preference for the body of the queen. Although originally called a louse it is really a tiny fly which usually occupies the regions of the bee's thorax lying between the bases of the wings and legs. As many as fifty or more have been taken at one time from a heavily infested queen. This insect does not suck the blood of its host as some have supposed, but instead it steals some of her food as she is receiving it from the nurse-bees. Since the queen is fed frequently and liberally it is not surprising that the Braula cocca breeds and multiplies more rapidly on her than on a worker bee.

When present on a queen in large numbers the insects impede her movements but otherwise they appear to cause her no physical injury. They hold tenaciously to their host when disturbed, but can be dislodged by subjecting the queen to a puff of tobacco smoke in a closed match-box.

Illingworth (40, p. *4*), confirming an old German practice, has found that small pieces of camphor, strewn on a sheet of strong brown paper spread on the hive-floor of an infested stock for one or two days, causes the insects to fall from the bees in a dying condition on to the paper. From this they can be removed and destroyed.

A full description of this creature and its mode of life is given by Herrod-Hempsall (33, p. *1484*).

It lays its eggs under the surface of the honey cappings (33, p. *216*). From these emerge larvae which burrow tunnels through the cappings, leaving holes when they emerge as pupae.

Fortunately this pest has become rare in Great Britain where at one time it was fairly common.

PLATE VII

THE QUEEN [Photo: L.E.S.

CHAPTER IV

BROOD FOOD AND ROYAL JELLY

MOST writers on the natural history of bees refer indifferently to the food supplied to worker larvae during the first two days of their life, and to that subsequently given to those larvae destined to become queens, as one and the same substance under the name "royal jelly". That they are not the same was demonstrated by Werner von Rhein (90, p. *664*) in his experiments on female dimorphism in hive bees.

Von Rhein reared worker and queen larvae without the help of hive bees, by keeping them in a suitable incubator. By means of a small syringe, to which a fine pipette was attached by rubber tubing, he withdrew larval food from the cells of a natural colony of bees and administered it to the larvae in the incubator,

He first fed 24 young worker larvae throughout the whole period of larval life with food taken from the cells of other larvae under two and a half days old. Of these only 5 reached or exceeded the weight of normal full-grown worker larvae and only one got as far as to be about to pupate. He concluded therefore that the food of young worker larvae alone was inadequate for rearing either workers or queens.

He then fed 28 two- to three-day-old worker larvae in queen-cells on royal jelly taken from other queen-cells, adding fresh jelly frequently because of its tendency to dry. Nine of these larvae pupated but all turned out to be workers, not queens.

The fact that he succesfully reared larvae on royal jelly but could not do so on worker-brood food proved that these foods are not identical.

In further experiments von Rhein fed larvae entirely

on large quantities of the mixed food, containing pollen and honey, which is found in the cells of older worker-larvae. Extra large workers were produced with pollen baskets, etc., but their ovaries were also extra large and their spermathecae well developed. He considered they might be capable of mating and laying fertilised eggs, but he failed to get them accepted as queens when introduced to stocks.

Von Rhein's failure to raise queens by feeding young larvae solely with

(*a*) The food of young worker-larvae,
or (*b*) The food of older worker-larvae,
or (*c*) Royal jelly,

suggests the absence of some essential condition or factor, e.g., the day to day variation in the composition of the jelly, or as von Rhein assumed, a hormone-like constituent provided by the nurse-bees.

The most marked changes in the composition of queen-larval food occur on the first and second larval days (p. 66) and if, as we must assume, these changes are essential for the full development of a queen, von Rhein's two- to three-day-old larvae, having existed as worker-larvae during a critical nutritional period, experienced an unfavourable start in life, a disadvantage supplemented by possible effects of the artificial conditions in which they were subsequently reared.

Royal jelly.

Royal jelly as supplied to queen-larvae is a rich and concentrated food having the appearance and approximate consistency of condensed milk. It is acrid to the taste, dries quickly, even within the hive, unless frequently replenished by the bees, and when dry, as in vacated queen-cells, becomes a gluey yellowish semi-solid mass. If not allowed to dry it can be preserved unimpaired for months. A secretion of the pharyngeal glands of the worker bee, and given in abundant quantity to queen-larvae, it promotes rapid growth and a wide differentiation of organs, both external and internal, from those of the workers.

It is rich in certain vitamins (p. 73), owes its keeping properties to its sugar concentration and a germicidal constituent, and has been shown to contain a hormone-like substance (27, p. *782*) which may or may not influence the development of the generative system in the queen-larva. Like the larval food of workers it contains suspended pollen grains in varying amounts, the presence of which is considered by some investigators to be accidental and non-essential (27, p. *781*). In the specimens examined by the present writer many of the pollen grains were broken which suggests that they are added by the bees, but whether they are necessary as an adjunct to the secreted royal jelly for the nutrition of the queen-larva is not known. They are, however, always found in the intestinal tract of the larva (31, p. *1*).

Several investigators, notably von Planta (1888), Appeler (1922), Elser (1929), and Vivino (1942), have made analyses of royal jelly. We shall refer to these only in so far as they appear to be of significance to the queen-breeder. For fuller information concerning them the reader is referred to Professor Haydak's work and his comprehensive review of the whole subject entitled "Larval Food and the Development of Castes in the Honeybee".

Differences between the food of young queen-larvae and young worker-larvae.

Von Planta (1888) first showed that royal jelly differs in composition from the food of worker-larvae under four days old. His figures for the percentage of the principal constituents were:

	Proteins	Fats	Invert sugar
Queen larva . .	45·14	13·55	20·39
Young worker larva	53·35	8·38	18·09

Vivino, collaborating with Haydak, analysed samples of royal jelly and young brood-food carefully collected by Haydak (28, p. *68*) with the following results:

	Age of larva in days	Proteins	Fats
Queen larva .	1	40·43	7·59
	2	48·85	5·61
Worker larva .	1–2	78·33	17·70

These results show that at the end of the first day the food of the queen-larva is considerably less rich in proteins and fats than that of the worker larva of the same age.

Vivino (28, p. *69*) found also that the average acidity of the food of worker larvae of 3 to 5 days of age (pH=3·9) was greater than that of queen larvae of the same ages, (pII–4·18).

Maisonneuve (62, p. *72*) citing Zander and Meier, states that these investigators found that the growth of the ovaries of a worker larva was considerably retarded from the end of the first larval day in comparison with that of a queen larva of the same age, and they concluded that a perfect queen could not therefore be reared from a worker larva more than one day old.

From these considerations therefore we must conclude that in order to obtain the best results in queen-rearing we should breed from eggs, as described in the "BHS" method (p. 265) or at least from larvae not more than one day old.

It may be stated here that one-day old larvae are very small—about the size of a pin's head—and that not only is it somewhat difficult to transfer them successfully to artificial cells, but the bees, probably on account of this, accept them less readily than older ones. It is for these reasons that several writers advise the grafting of larvae up to 36 hours of age. It is preferable however to graft with the younger larvae and to risk a proportion of rejections.

In Pl. VIII larvae of all ages are shown. A few one-day old larvae are seen near the centre of the picture

Changes in the composition of the larval food of workers and queens.

The following extract from Vivino's table of results (27, p. *780*) of the analysis of samples of food taken from larvae of different ages serves to show the variations in the composition of royal jelly as compared with those in the food of worker larvae.

Percentages of dry matter.

	Age of larva	Proteins	Fats	Ash
Royal jelly.				
	1	40·43	7·59	3·34
	2	48·85	5·61	2·95
	3	50·63	16·13	2·63
	4	46·20	18·74	2·31
	5	49·75	15·18	2·34
	Sealed	58·01	12·59	2·37
Larval food of workers				
	1–2	78·33	17·70	4·04
	3–5	50·39	5·87	1·65

The table shows:

(1) *Protein content*

(*a*) Royal jelly. Considerable increases on the second and sixth days. An average of nearly 49 per cent for the intervening days,—the main growth period.

PLATE VIII

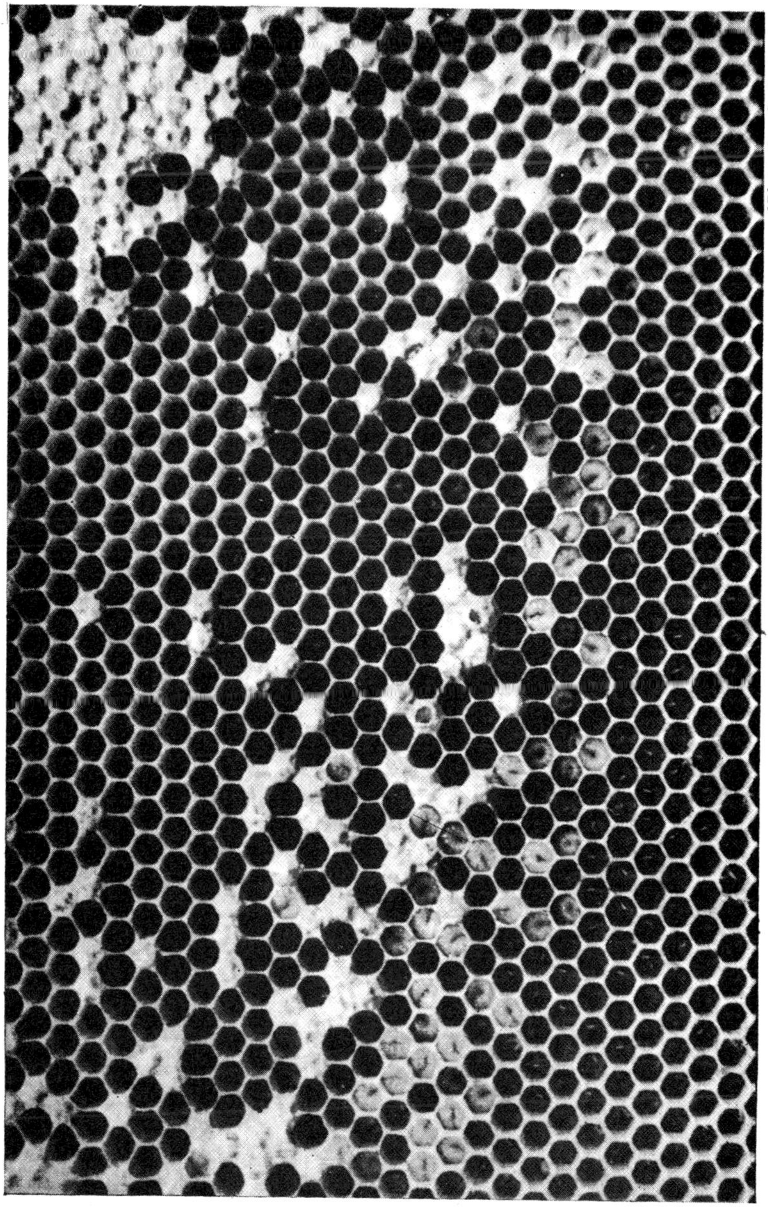

[Photo: Rev. G. H. Hewison

EGGS AND BROOD IN VARIOUS STAGES. ONE-DAY OLD LARVAE (SUITABLE FOR GRAFTING) NEAR THE CENTRE (*b*) Worker-larval food. A very high initial content which drops after the second day to approximately the same percentage as for queen larvae of the same ages.

(2) *Fat Content.*

(*a*) Royal jelly. A decrease after the first day followed by an approximately three-fold increase for the third to the fifth day inclusive. A decrease on the sixth day.

(*b*) Worker-larval food. A high initial content, reduced to about one-third for the ensuing days.

(3) *Ash* (*minerals*).

(*a*) Royal jelly. A moderate reduction on the second day.

(*b*) Worker-larval food. A heavy reduction after the second day. The average percentage of mineral matter (ash) in royal jelly is about one and a half times that in worker-larval food for the main growth period.

Thus we note that

(1) There are definite changes in the content of protein, fat, and minerals in the food of the queen-larva at the end of one day.

(2) The worker larva of the first two days receives a food much richer in protein, fat, and minerals than that of the queen larva of the same age, but the proportions of these substances in the case of the worker larva are greatly diminished after the second day.

(3) During the main growth period—third to fifth day inclusive—the food of the queen-larva contains about three times as much fat and about one and a half times as much mineral matter as that of the worker larva.

Use of added royal jelly in grafting.

In view of the day to day variation in the composition of royal jelly the queen-breeder might naturally enquire whether royal jelly taken from queen-larvae of one age is suitable for grafting larvae of another age (p. 238). In particular he might suspect that the use of royal jelly taken from larvae of advanced age would be unfavourable to the acceptance by the bees of younger grafted larvae.

To decide this point the present writer carried out the following experiment:

Experiment with royal jelly.

Twenty-four artificial cells were grafted by the Doolittle method (p. 241) with larvae approximately one day old. The wooden cell-holders were numbered in threes,—1, 2, 3, 1, 2, 3, and so on. The royal jelly supplied to all the cells numbered 1 was taken from queen-larvae approximately 30 hours old; to all numbered 2 from queen-larvae of about 75 hours; and to all numbered 3 from queen-larvae of approximately 96 hours of age.

The three lots of royal jelly were taken from the respective larvae on the day of grafting and were removed as needed directly from the queen-cells.

To distribute fairly any possible effects of drying or chilling the cells were grafted consecutively, viz:—1, 2, 3, 1, 2, 3, and so on.

The grafted cells were given to queenless bees confined in a swarm box. Seventeen out of the twenty-four were accepted.

It will be noted from the illustration (Pl. IX) that the acceptances include ten which are consecutive in the upper row. Of the rejections two are numbered 1, two numbered 2, and three numbered 3. The curious rejection of five consecutive cells in the lower row cannot be explained, but this does not impair the experiment.

PLATE IX

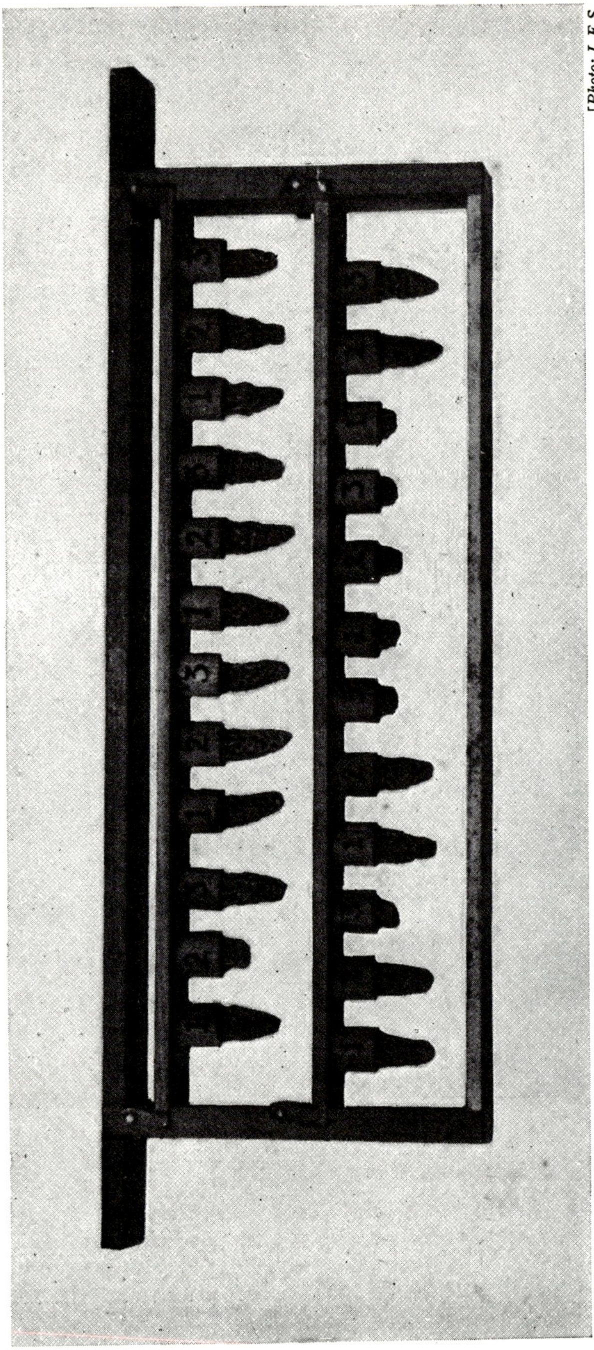

[Photo: L.E.S.

RESULT OF GRAFTING WITH ROYAL JELLY TAKEN FROM LARVAE OF THREE DIFFERENT AGES

The results afford no evidence that the age of royal jelly, taken fresh from unsealed queen-cells and used in grafting, is of any practical significance in relation to acceptance by the bees.

Interruption of larval feeding.

Bees are partial to royal jelly and devour it greedily when it is separated from a larva. They sometimes,—if not always—consume the jelly provided by the queen-breeder in grafting within a few minutes and then provide a new supply for the larvae they accept. In such cases the added jelly at least serves to prevent drying and starving of the young larvae until they receive their appropriate food. This interruption of natural feeding however is avoided in the process known as "double-grafting" (p. 242).

In this connection we may note that some queen-breeders graft "dry", that is to say without added royal jelly. In this case the point of the grafting tool is carefully placed under a larva so as to take up as much as possible of its small supply of food. If this is successfully done the larva is accepted as readily as one provided with added royal jelly.

Haydak (27, p. *753*) expresses the view

> "that the production of either a worker or queen is due not to the change of food but to the amount of essential nutrients consumed by the queen and worker larvae. Up to about the third day of age, all female bee larvae are in the period of mass feeding (Lineburg, 1924). After that time the larvae in the worker cells are fed progressively, the food being given to them in the intervals as needed. While the queen larvae continue to be on a mass feeding basis, the worker larvae are undernourished. 'The queen and worker larvae have the same approximate growth rate during early larval life and then the growth of the worker caste is retarded (Melampy et al., 1940).'"

He goes on to point out that the worker larva is starved after sealing and has to spin its cocoon at the expense of its own body, whilst the queen-larva, provided with a reserve of food, grows and increases considerably in weight after sealing, notwithstanding the expenditure of materials used in the spinning of the cocoon. He continues,—

> "So in the case of the queen-larva there is no underfeeding at any time of larval life, while in the case of a worker-larva a partial inanition starts after the larva attains the age of two to five days."

By removing larvae from sealed queen-cells and keeping them in suitable conditions of temperature and humidity, but without food, Haydak (27, p. *785*) was able to show that the younger larvae, averaging 204 mg. in weight, developed into worker-like insects, only one of which became a live adult, whilst the older ones, averaging 232 mg. in weight, developed into queens. He thus showed that even at this late (pre-pupal) stage, total inanition may prevent the normal development of a queen.

Hence we realise the importance of rejecting queen-larvae if they have no reserve of food in their cells before sealing. It is commonly known that queens reared in weak and hungry stocks have little or no reserve of food, especially before sealing, and that whilst they may mate and lay fertilised eggs they are undersized, and often capable of passing through queen excluders with unfortunate results. Such queens have short lives and are liable to early supersedure.

VITAMINS

Until the first decade of the present century it was generally considered that only proteins, carbohydrates, fats, minerals, and water in suitable forms and proportions, were necessary for the nutrition of animals.

It was then found by experiments on small animals that balanced diets of purified artificial foods, containing these substances only, failed to maintain growth, and induced serious deficiency diseases which resulted in death. Diseased animals were rapidly restored to health by the addition to their diets of small quantities of certain natural foods, and investigators therefore concluded—what they had long suspected—that these natural foods contained minute quantities of unknown substances which were essential to the maintenance of growth and health in animal life. These unknown substances were at first called "accessory food factors" and later "vitamins".

During the last 30 years a considerable number of these essential food factors have been discovered, their biological activities ascertained, the chemical composition of many determined, and a number of them have been synthesised. Their discovery has led to a revolution in the science of nutrition and the laws of health.

Originally the vitamins were denoted by the letters of the alphabet, A, B, C, D, E, etc. Later, Vitamin B was found to be a mixture of several substances and these were called B_1, B_2, etc. These were also given their chemical names. At present both systems are used, some vitamins retaining their literal symbols whilst others are known by their chemical names.

Some of the vitamins are soluble in fats and others in water. The fat soluble vitamins, A, D, and E are absent or are found only in negligible traces in royal jelly. Of the water soluble vitamins B and C the latter is found only in traces, but several of the complex of B vitamins are present in significant quantities. To the importance of these in animal nutrition we can refer only in the briefest possible way:—

Thiamine, (B_1). Associated with the metabolism of carbohydrate in the nerve-cells. Necessary for the normal activity of the central nervous system.

Riboflavine, (B_2). Essential for growth, the health of the skin, and the oxidation of foodstuffs.
Pyridoxine, (B_6). Protects against certain skin diseases.
Nicotinic Acid, (B_7). Prevents the disease "Pellagra".
Pantothenic Acid. Present in all animal tissues.
Biotin. Growth promoting.
Inositol. Growth promoting.
Folic Acid. An essential growth factor.

These vitamins are not all or equally necessary in the nutrition of different animals. The specific value of each in the nutrition of the queen-bee larva has not been determined.

For a fuller account of the vitamins in respect of animal nutrition generally the reader is referred to Heilbrunn's "An Outline of General Physiology" (32, Chapter 15).

Vitamin content of royal jelly.

The presence of vitamins in royal jelly has been determined by several investigators (27, pp. *782–3*). Their results, mostly ascertained by microbiological methods, vary considerably.

Appeler (2, pp. *151–153*) reported little or no vitamin A, but a high content of B vitamins in royal jelly.

At one time it was thought that the presence of vitamin E, which promotes fertility, was responsible for the transformation of a worker-larva into a queen-larva, but recent investigators including Mason and Melampy, and Haydak and Palmer (27, p. *782*) have found no evidence of its presence. Ascorbic acid (vitamin C) has been detected in royal jelly only in traces, and no vitamin K activity (responsible for the clotting of blood) has been found. (28, p. *70*).

It is in vitamins of the B complex that royal jelly is specially rich. Haydak (28, p. *70*) gives the particulars of these as follows:—

"One gram of fresh royal jelly contains from 3 to 18 micrograms[1] of thiamine chloride, (B_1); from 8 to 28 micrograms of riboflavin, (B_2); from 2 to 50 micrograms of pyridoxine, (B_6); from 59 to 111 micrograms of nicotinic acid; 1·6 to 4·6 micrograms of biotin; from 78 to 150 micrograms of inositol; and from 0·2 to 0·5 micrograms of folic acid. . . . Royal jelly is especially rich in pantothenic acid, (up to 320 micrograms per gram). It seems to be the richest source of this vitamin."

Pearson and Burgin (58, pp. *415–417*) of the Agricultural and Technical College of Texas, in 1941 reported results of their investigation of the pantothenic acid content of royal jelly. Using the microbiological method of assay on freshly collected material they found from an examination of seven samples that it contained from 142 μg. to 200 μg. (average 183 μg.) on a fresh basis, and from 393 μg. to 580 μg. (average 511 μg.) of pantothenic acid per gram on a dry basis (i.e., after abstraction of water), amounts which are considerably higher than for any other known material.

Two of the substances richest in this vitamin are yeast and liver, the average for dried brewer's yeast, according to Jukes, cited by Pearson and Burgin (58), being 200 μg., and for liver 180 μg. per g.

On this basis the ratio of the amount of pantothenic acid in royal jelly to that contained in yeast is more than $2\frac{1}{2}$: 1. Microbiological methods give considerably lower values for yeast and liver and in comparison with these the ratio of the amount of the vitamin found by Pearson and Burgin in royal jelly to that found in yeast is 6 : 1.

These investigators conclude that royal jelly is the richest known source of pantothenic acid and that on a dry basis it contains from 2·5 to 6 times as much as

[1] A microgram (μg.) is one millionth of a gram. One gram is approximately one twenty-eighth of an ounce.

is contained in yeast or liver. They add the following comment:—

> "Of the vitamins studied in royal jelly pantothenic acid is the only one in which it is unusually rich. Whether or not pantothenic acid is in any way responsible for the development of female larvae into queen-bees remains to be studied. The presence of pantothenic acid in such relatively large amounts does suggest that it might be one of the factors responsible for the development into queens of the female larvae which continue to receive royal jelly".

Origin of vitamins in royal jelly.

Royal jelly must be derived from honey and pollen, the only foods of the bee. Vitamins in various American honeys have been determined by Haydak and others (29, p. *586*) who found them to contain minute and varying quantities of thiamine, riboflavine, pyridoxine, pantothenic acid, nicotinic acid, and ascorbic acid. By filtering with diatomaceous earth they found the vitamin concentration was reduced by from one-third to almost one-half of the original values. They attributed the diminution as partly due to the removal of suspended pollen.

Kitzes and others (42, p. *420*) observe that these vitamins exist in honey only in fractions of the amounts necessary for the metabolism of the honeys in which they are found.

It cannot be assumed therefore that the vitamins in royal jelly are derived in any appreciable measure, if at all, from honey.

Pollen however contains all the vitamins found in royal jelly in amounts which vary from season to season and with the different plants from which it is derived. Haydak and Palmer (30, p. *320*) using bee-bread (stored pollen) and royal jelly, both of the 1940 crop found that the former contained more thiamine, riboflavine, pyridoxine, and ascorbic acid than royal jelly, but less of nicotinic acid and pantothenic acid.

Pearson (59, p. *291–292*) using samples of fresh pollen, most of which were mixed, and secured from bees by pollen-traps, found an average pantothenic acid content of 30·3 μg. per gram of dry material. On the basis that royal jelly contains 511 μg. per g. (p. 75) he concluded that it contains about 17 times as much of this vitamin as the average for pollen. "This indicates", observes Pearson, "that the honey-bee either has the ability to synthesize pantothenic acid or that there is a marked concentration of this vitamin which is transmitted to the royal jelly". He goes on to say that of the B vitamins studied pantothenic acid is the only one in which the amounts in royal jelly and pollen are not of approximately the same order. This, he concludes, affords support for the suggestion that this vitamin may be one of the factors responsible for the phenomenal properties of royal jelly in developing female larvae into queen-bees.

Pollen substitutes.

In some countries (not usually in England) there are periods during which the supply of natural pollen is deficient and certain pollen substitutes are recommended for use. Haydak and Tanquary (31, p. *13*) have tested the comparative values of these, both in relation to ordinary brood-rearing and to queen-rearing. Small colonies of bees, deprived of natural pollen from the time of their emergence, were caused to raise queens whilst being fed on honey and various pollen substitutes. The queens were weighed in each case seven days after emergence from their cells.

The heaviest queens were produced in stocks fed with a mixture of honey and a pollen substitute comprising dry skim milk with 20 per cent dried non-irradiated yeast, in the proportions of 200 g. honey and 35 g. substitute.

With a pollen substitute consisting of dried skim milk alone (which would be deficient in fat) no queens were reared.

Incidentally we may here remark that of the 24 separate substances used as pollen substitutes by these investigators, pea-flour, for many years popularly recommended as a pollen supplement in England, and still so used by some people, holds a very low place in their table of values. It induced the highest rate of mortality and inhibited all brood rearing.

Since, as the foregoing considerations indicate, the vitamins in royal jelly are derived from pollen, the provision of an abundant supply of pollen for queen-raising bees is of great importance. That bees attempt to raise queens when deprived of it can be demonstrated, as in the writer's experiment described below, but this they do at the expense of their own bodies, the queen-cells being small and the inmates failing to reach maturity if additional brood has to be cared for.

Attempt to rear queens by bees deprived of pollen.

In July of the present year the writer caused bees, provided with abundant honey but deprived of pollen, to attempt to raise queens. The hive selected comprised an ordinary brood-box and two shallow-comb supers above an excluder, the upper super being filled with freshly gathered honey, almost entirely sealed. No pollen cells could be found in this super which was fairly well crowded with inactive bees.

A screen of perforated zinc was placed between the two supers thus imprisoning and rendering queenless all the bees in the upper super. The conditions of these bees therefore were that they shared the air and temperature of the stock below, and were provided with an abundance of honey but no pollen.

A comb containing a large patch of eggs on each side, most of which had hatched, or were on the point of hatching, was then inserted in the middle of the

PLATE X

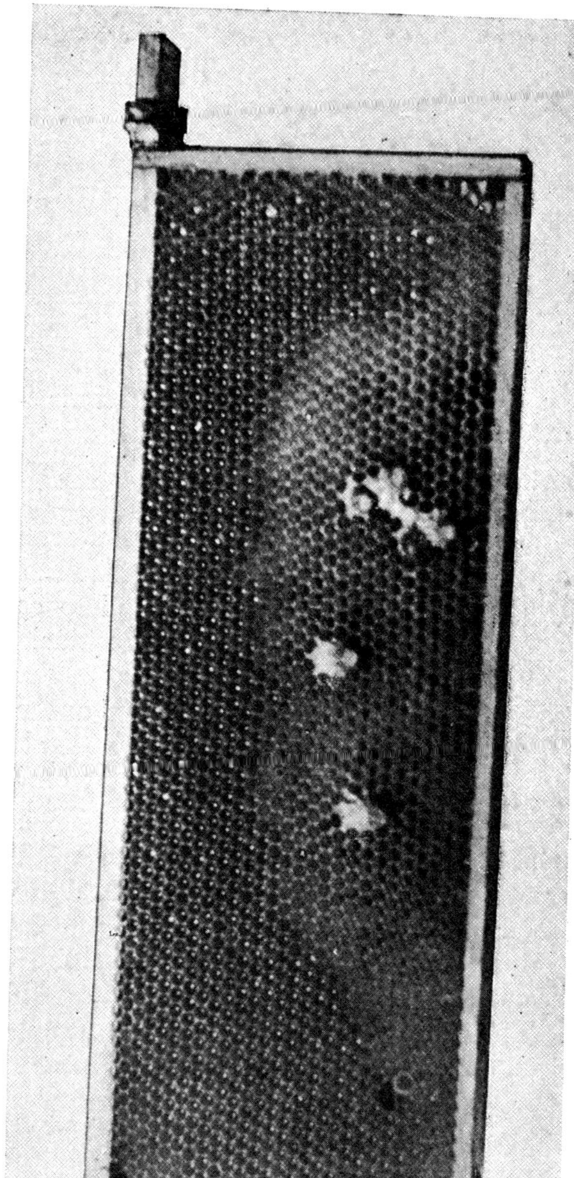

[*Photo: L.E.S.*

QUEEN-CELLS RAISED BY BEES DEPRIVED OF POLLEN

super, and the imprisoned bees left undisturbed for five days.

When examined on the fifth day the comb had the appearance shown in Pl. X. Six small queen-cells had been built and sealed. With the exception of one rather larger than the others, none projected more than $\frac{1}{4}$ inch beyond the face of the comb.

The surrounding brood, which under normal conditions would have been sealed at about the same time as the queen-cells, was open, and remained open for a further two days. A patch of brood about the size of the palm of one's hand had by that time become sealed.

Four days later the queen-cells were cut out, placed in cell protectors, and transferred to nuclei prepared to receive them. Three were destroyed by the bees at once, presumably because there were no protective cocoons (p. 38), and the remaining three failed to yield queens. When examined these were found to contain dead queen pupae without a trace of residual royal jelly.

The task of providing royal jelly for six queen-cells and brood-food for the surrounding brood had been beyond the capacity of the bees. Dependent on honey and their bodily reserves for the ability to secrete brood-food they had concentrated on the feeding of the queen-larvae, which, inadequately nourished, they sealed over at the normal time.

The surrounding brood was retarded by relative starvation and the date of its sealing was consequently deferred. The greater portion subsequently died and was removed by the bees without being first sealed over.

The bees used, being inactive occupants of sealed honey combs remote from the brood-nest, would not be likely to include any active nurse-bees. Their failure to raise queens successfully could not be attributed entirely to this circumstance for it is commonly known that bees of any age can, in suitable conditions—including access to both honey and pollen—raise queens to maturity. Their failure to do so in this case therefore appeared to be attributable to partial starvation caused by the absence of pollen.

Protein content of royal jelly.

We have already noted (p. 65) that von Planta found that the average protein content of royal jelly (dry matter) was 45·14 per cent. Vivino (p. 66) found that the average percentage for 7 daily stages of larval life (inclusive of sealed) was approximately 50. This proportion is very high compared with the content of human foods richest in this substance e.g., cheese, about 26 per cent.

Protein is the generic name for a group of highly complex chemical substances. It constitutes the most important part of the protoplasm of the body cells of all animals and occurs in considerable amounts in plants,—especially in seeds. All the proteins contain nitrogen and are therefore in nutrition described as nitrogenous foods. Some of the protein in a living organism is burnt up to produce energy.

In the course of digestion the proteins are hydrolised by enzymes, yielding simpler bodies called amino-acids. These are carried by the blood stream to all parts of the body, the various tissues abstracting from the blood the particular amino-acids they need. Some are essential for the building up and repair of tissues, some for the growth of young animals, others for the elaboration of internal secretions, including enzymes and hormones, whilst some are resynthesised into proteins.

Certain of these essential amino-acids have been identified in royal jelly (27, p. *781*).

Fat content of royal jelly.

Fats are important constituents of animal food. They are used as fuel by the tissues and in their intracellular forms are essential for the life of every body-cell. When emulsified in the course of digestion they are attacked by enzymes which convert them into fatty acids and glycerol in which forms they are absorbed into the blood stream. Some of these are later resynthesised into fats, and additional fat is synthesised from carbohydrates (sugars) and protein. Stored in the body they constitute reserves of fuel and form soft protective and heat-conserving curtains or padding around and between the internal organs.

As shown on p. 66, royal jelly, after the second larval day contains about three times as much fat as the food given to workers of the same age, or to put it in another way the partial starvation and retardation of the worker larva is partly due to a heavy reduction of the proportion of fats in its diet.

The fat cells in the "fat body" of the bee-larva have two important functions. In addition to the fat they contain they store large quantities of glycogen derived from the sugar content of the larval food and also accumulate quantities of protein granules. During the pupal period most of the cells disintegrate and liberate their stores of fat, glycogen, and protein into the blood plasma whence they are utilised for the nourishment of the growing tissues of the pupa and young imago (80, Chapter IX).

The bee-larva therefore absorbs food not only for its own nutrition but also for that of the subsequent non-feeding stages, the pupa, and the imago before its emergence from the cell.

Sugar content of royal jelly.

According to Appeler (2, pp. *151–153*) royal jelly contains approximately 15 per cent of sugars. These are comprised mainly of invert sugars (dextrose and levulose) which can be directly absorbed into the blood stream. In the bee-larva they are largely converted into glycogen which is stored up in the fat cells, and subsequently reconverted into the sugars before being burnt up in the tissues to provide heat and energy.

Effect of diet restriction on worker-larvae.

In a fertilised egg the course towards one sex is set by the chromosomes. Reasoning from the analogy of other animals, including some insects, we may assume that in the later post-egg stages of the female bee the growth of various organs, including the sex organs, is controlled by internal secretions which are poured into the blood stream by various glands. These including the sex glands themselves, are known as endocrine glands and their secretions as hormones. The growth of an organ, and its activity, are promoted or retarded by excess or deficiency of the particular hormone involved.

The supply of these secretions depends on adequate nutrition, including sufficient supplies of carbohydrates (sugars), amino-acids derived from protein, fats, minerals and vitamins. These are all present in royal jelly and being provided in liberal quantities to queen-larvae are conducive to the normal secretions of hormones and the complete development of the female bee.

The diet of the worker-larva, however, greatly restricted in quantity, and in comparison with that of the queen-larva, deficient in fats and minerals, may reduce or prevent the secretion of the sex hormones and so inhibit the full development of the generative organs and their associated sex characteristics.

But for this restricted diet, therefore, all female bee-larvae would normally become fully developed queens.

Age as affecting queen-larvae and nurse bees in queen-rearing.

In practice it is found that reasonably good queens can be reared from worker larvae of 36 hours of age and these are preferred by many who practise grafting on account of their increased size. Becker (43, p. *81*) ascertained that the limit of age of a worker larva from which a queen can be raised is from 3 days 6 hours to 3 days 18 hours. Periods in excess of 3 days 6 hours, noted by Becker, were accounted for by Komarov as due to retardation of development on account of less favourable circumstances of temperature and nutrition.

Becker found that queens reared from larvae of $3\frac{1}{2}$ days of age as compared with those from larvae of 3 days, showed a marked diminution of the number of ovarian tubes and of the diameters of their spermathecae, and Komarov noted in addition a diminution of the weights of the queens, of the lengths of the egg tubes, and of the diameters of the ovaries.

Komarov (43, p. *81*) noted also that queens raised from larvae up to 3 days of age usually develop normally in respect of their physical characteristics but he suggests that sharp variations in nutrition which many of the larvae undergo account for differences in quality of the resulting queens. He recommends that larvac aged from 12 to 24 hours be selected for queen-rearing because at this age they are more acceptable to the bees and the period during which they remain "in the condition" of future worker-bees is shortened.

Komarov further found that the ages of the nurse bees affected the quality of the queens reared by them. The older the bees the fewer were the queen-cells and the less the weights of the emerging queens. This was specially evident when successive batches of prepared cells were given to the same stock. The queen-cells were cut out as soon as sealed and fresh larvae given and this was repeated until the bees refused to accept the new larvae. The number of queen-cells raised, and the weights of the queens, diminished with each succeeding generation until a queen of the sixth generation was lighter in weight than a worker bee, which in some respects it resembled.

In the course of his experiments which were carried out with accurate determination of the ages of the nursing bees Komarov observed that:—

(1) Very young bees, being immature, reared few queen-cells.

(2) Hive (non-flying) bees selected young larvae from which to raise queens whilst older bees were not so particular.
(3) Bees may be old according to the calendar but may, on account of reduced activity, be physiologically young and therefore able to secrete brood-food.
(4) The greater the number of queen-cells reared at one time the lighter in weight will be the queens, the smaller their ovaries, and the less the number of their egg-tubes.

In normal conditions, according to Rösch (73, p. 5), the pharyngeal glands of the worker, which secrete brood food and royal jelly, are in a state of maximum efficiency when the bee is from 6 to 15 days old. Bees of this age are properly called "nurse" bees although older bees are capable of fulfilling their functions.

In the light of this and Komarov's valuable observations we realise why:—

(1) We should use bees taken directly from the brood combs for rearing queens.
(2) The less the number of queens raised at one time the better they will be.
(3) Queens raised under the supersedure impulse, when very few queen-cells are made, are often excellent.
(4) It is not desirable to raise successive batches of queens from the same stock unless there are intervals of time between them.

CHAPTER V

DRONES

Life history of the drone.

As explained in Chapter VII drones are produced from unfertilised eggs. These are deposited by the queen in the drone cells, of which the normal hive contains a number—usually a few hundred. Having to accommodate a larger insect these are larger than the cells in which workers are bred, being $\frac{1}{4}''$ in diameter as compared with $\frac{1}{5}''$ diameter of the worker cell. Their depth too, is greater, being $\frac{5}{8}''$ as compared with the $\frac{1}{2}''$ of the worker cell. The greater depth is partly secured by raised domed cappings, the cappings of worker cells being approximately flat.

The thickness of a comb containing sealed drone brood on both sides is $1\frac{1}{4}''$ whilst that of a comb similarly provided with worker brood is slightly under $1''$. This is a point of some practical importance for the amount of drone brood raised in a hive can be restricted by placing the combs so near together that the bees cannot complete drone cells to their proper depth.

In a normal stock the queen usually deposits a few eggs in drone cells at the opening of the season, that is, when the combs are well stocked with worker brood and nectar and pollen begin to be brought in. In this country this generally happens at the end of March or early in April and it is the first step the bees take in preparation for swarming. The resulting drones are not available for mating until five or six weeks later—about the time when natural swarming may take place and young queens are soon to be expected. Exactly what induces the queen to lay a proportion of drone eggs during the period when she is mainly engaged in producing large quantities of worker brood is not known.

The drone takes a longer time to reach maturity than either the queen or worker. Its life-stages, as compared with those of the worker, are tabulated by Cowan (20, p. *12*) as follow:—

	Worker *Days*	*Drone* *Days*
Incubation of egg	3	3
Feeding of larva	5	6
Spinning of cocoon by larva	2	3
Period of rest	3	4
Transformation of larva into pupa	1	1
Time in nymph state	7	7
Total	21	24
	Day	*Day*
The cell is sealed over	9th	9th
The bee leaves the hive to fly	14th	14th

It will be noted that the drone spends three days more than the worker in the larval stage and therefore emerges from its cell in 24 days instead of 21 days after the laying of the egg. It is supplied with abundant food but as in the case of the worker is "weaned" on the fourth day of the larval stage by the addition of honey and pollen to the secreted brood food.

Before it reaches sexual maturity the adult drone passes through a period of physical development. Root (72, p. *225*) citing Bishop, states that "it undergoes a further growth period of at least nine to twelve days. During this period many changes take place in the drone organs, especially up to the sixth day, after which the changes are slight". As it does not usually fly until fourteen days old it is not considered sexually competent until it reaches that age. The point of all this is that in the business of queen-rearing we should take the first step in the provision of drones about seven weeks before our young queens are expected to take their mating flights or what is the same thing, a month before we select eggs or larvae from which to rear queens.

The adult drones feed on honey and pollen. Whilst maturing during their first twelve days they probably receive also some brood-food from the nurse bees (p. 87).

They are plentiful from the time of swarming until the end of the main honey harvest. When the latter ceases and the bees are faced with the need of economy, and when in the ordinary course of nature there are no more young queens to be mated, the drones, which do not contribute to the labour of the hive, are deprived of brood-food. For a time they congregate in one part of the hive—often on the comb most remote from the entrance—and finally they are seized, bitten at the base of the wings by the workers, and expelled from the hives to die. The "massacre" of the drones, usually in August, is a familiar and pathetic sight.

Mating of drones.

During the summer months the drones fly freely on warm afternoons and to a more limited extent during other parts of the day. Those that have completed their early practice flights are available for mating (p. 87). They probably continue to be sexually potent as long as they are able to fly vigorously.

It is here unnecessary, and beyond the scope of this work, to describe in detail the generative organs of the drone. They are shown in the illustrations in Plate XI, but for a full description of them the reader should consult a work on bee anatomy. Briefly they comprise:—

(1) two testes in which the spermatozoa are developed during the larval and pupal stages of the drone's

PLATE XI

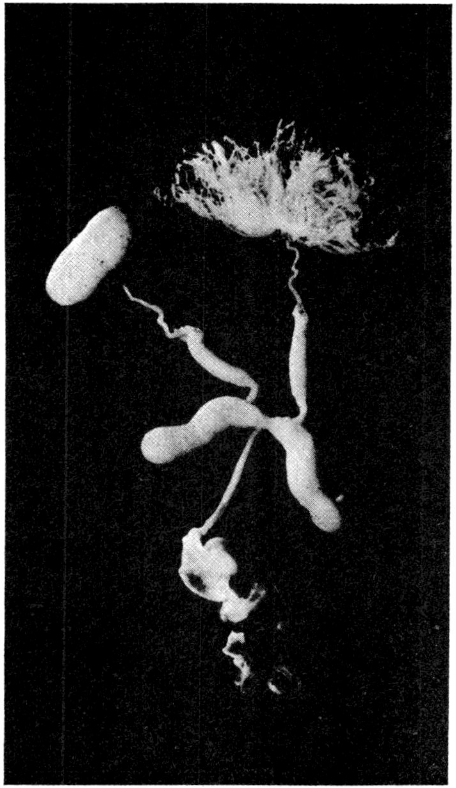

[*Photos: R. Bassindale*

GENERATIVE SYSTEM OF DRONE

A. Of young drone. *B*. Of old drone.

T. Testes; *V D*. Vasa deferentia; *V S*. Vesiculae seminales; *M G*. Mucus glands; *D E*. Ductus ejaculatorius; *B P*. Bulb of penis.

life. Each testis comprises a large number of fine tubules in which the spermatozoa are formed and develop. These tubules are shown in the teased out testis of a young drone in Pl. XI, A.

(2) the coiled tubes (*vasa deferentia*) which convey the sperms from the testes to two spermatic reservoirs (*vesiculae seminales*). In these reservoirs the sperms reach maturity.

(3) two short ducts connecting these reservoirs with the bases of two large mucus-secreting glands, and emerging from near the joint base of these glands one long tube (*ductus ejaculatorius*) along which both sperms and mucus fluid are forced to the bulb of the penis. In an older drone the testes are shrunken (Pl. XI, B) the spermatozoa having left them and accumulated in the bulb.

(4) The penis—a large, irregular shaped organ of which the most important parts are:—

  (*a*) a bulbous enlargement (the "bean") in which the sperm is collected before coition, and

  (*b*) two horn-shaped appendages (*pneumophyses*) which at the time of coition are inflated, and fitting into corresponding recesses in the vagina of the queen, facilitate the entry of the unfolding penis and possibly cause it to be retained in position when it is separated from the drone (p. 91).

The sexual organs of a live drone are extruded, and may be examined, if the insect be chloroformed, or if the abdomen be gently pressed between the thumb and finger. A drone which has recently died may be used for the latter treatment.

Although many queens mate more than once (pp. 45–47) a single mating is usually sufficient for the whole period of a queen's fertility.

Copulation takes place in the air. It was formerly supposed that the queen flew high and to a considerable distance and was chased by numbers of drones, the fastest, and therefore the most vigorous of which, overtook her. It is now well established that the union may take place within a short distance of the hive. It is not commonly observed, but has been described by eye witnesses who agree that the insects unite in the air and after a short interval of spinning flight fall to the ground before separation. Spiller (83, p. *46*) who has observed the mating of hive-bees and also that of solitary burrowing bees of the genus Halictus states that the mode of union is the same in both cases, i.e., that the two insects unite in the form of letter S, bringing the tips of their abdomens together between their feet. The relative positions of the two bees are illustrated in a drawing by Spiller which appeared in "The Bee World" of November, 1935.

The end of the union is marked by a light explosive sound which Caird (16, p. *99*) compares with that resulting from the bursting of a seed pod in the heat of the sun. This, he states, marks the rupture of the terminal part of the drone organ and the ejection from it of the seminal fluid and subsequently the mucus.

The queen disengages and returns to her hive carrying with her the ruptured drone organ the protruding part of which Caird describes as a "glistening white thread, about $\frac{1}{8}$ in. in length, consisting of the collapsed terminal portion of the drone's organ, filled with mucus".

The discharge of spermatic fluid takes place near the entrance to the spermatheca (p. 52) some being forced into the lower ends of the paired oviducts (Pl. V, OvD). Within six and a half hours (80, p. *259*) it has practically all entered the spermatheca. A few hours after the mating the shrunken male organ comes away or is removed by the bees.

The drone dies at the end of copulation or immediately afterwards.

The appearance of the white organ of the drone protruding from the tip of the abdomen of a returning young queen is evidence that copulation has taken place.

Miss Betts (12, p. *142*) reviewing an article which appeared in Deutscher Imker, 1934, cites the experience of F. Kaspar who observed that in the mating flight the queen's wings remained folded all the time he kept the pair in sight. From this we may infer that the drone probably needs superior strength to support the weight of the queen.

In another connection Miss Betts expresses the opinion that an undersized drone (e.g., from a worker cell) whilst probably unable to support a normal-sized queen, may nevertheless mate successfully, unless it can be shown that mating *must* take place in the air and not merely just after violent flight.

For a detailed and illustrated description of the genital organs and the mating of the queen and drone the reader is referred to Herrod-Hempsall's "Bee-keeping, New and Old", II, pp. 1615–1635.

Influence of the drone in breeding.

In all forms of domestic stock-raising in which improvement is sought the qualities of both male and female parents are of the greatest importance. In the case of the commoner domestic animals,—e.g., the horse, cow, and sheep, where one male is mated to many females, the selection of the male is of even greater economic importance than that of the female. In a consideration of the influence of the male bee however the following factors are involved:—

(1) A drone mates only once and therefore in this respect its influence is limited. This limitation however is neutralised by the abundance of drones of the same kind normally produced in a single stock.

(2) Being produced parthenogenetically (p. 124) the drone has no father. Its hereditary characteristics are therefore those of its queen mother, which in turn are those of her father and mother; or we may say that a drone inherits the qualities of its grandfather and grandmother which are combined in its queen mother. As we have no direct means of assessing the transmissible qualities of a drone, apart from physical characteristics such as colour, we consider it to transmit the qualities of its mother, or, more correctly, of its grandparents.

(3) Apart from isolation (p. 303 et seq.) and artificial insemination (pp. 50–51) mating cannot be certainly controlled. We have to judge of drone parentage by the characters of the progeny resulting from a mating, bearing in mind that where bees of different races are kept in the same neighbourhood there is a marked tendency to crossing (p. 46).

(4) Parthenogenesis is favourable to a change from one race to another. For example, suppose a beekeeper having only Italian bees in an isolated apiary wishes to change over to British "Blacks". He introduces a fertile black queen to one stock and from it rears sufficient queens for all the other stocks. Most of the queens, mated to Italian drones, will produce hybrid worker bees (p. 140) and black drones. A small proportion, mated with drones from the black stock will produce black workers and drones.

At this stage therefore all the drones in the apiary will be blacks. If now a fresh lot of young queens be raised from the original black mother and substituted for the hybrid queens, all the resulting progeny, in theory, will be blacks.

In practice complete isolation is rarely possible so

that if the race is to be kept pure the continued breeding from pure black queens and the elimination of occasional hybrids resulting from mis-mating, or from plural mating (p. 45) will be necessary.

(5) Similarly parthenogenesis tends to prevent the establishment of a fixed hybrid strain from male and female parents of different races.

Suppose, for example, we wish to obtain and perpetuate a hybrid strain from male and female parents of two distinct varieties,—e.g., Black (male) and Italian (female) as judged by colour, and let us assume that for our purpose a black stock is introduced to an isolated apiary of Italians.

To secure uniform crossing the black stock is encouraged to produce drones freely whilst drones of the Italian stocks are limited and prevented from flying (pp. 101–102).

Young queens are raised from a selected Italian stock and these, flying from drone-free nuclei, must mate with black drones. These young Italian queens are then substituted for all the original queens in the apiary. They will produce 100 per cent hybrid females (workers and queens) and the hybrid strain desired is temporarily established. But all the male progeny of this first generation of cross-mated queens will be Italians (p. 140 and Pl. XIII, B, F2).

Any union within the strain must now be between a hybrid queen and a pure Italian drone. Reversion to one of the pure races then begins for the female progeny will be half hybrids and half pure Italians, and the male progeny half Italians and half blacks (p. 143 and Pl. XIII, B, F2).

The future course of unrestricted breeding within the strain and the tendency to reversion to the pure types is discussed on pp. 144–148. There is therefore a tendency, based on Mendel's laws, for hybrids to revert to pure types. This tendency is promoted by the parthenogenetic inheritance of the drones which cannot be hybrids, and must be of one or other of the pure types.

Cheshire (17, II, p. *619*), referring to what he calls the "disposition to breed out the drone element", writes:—

"It is clear that the queen exerts three times the influence over her posterity that the drone does, since her blood is one-half of that of her fecundated daughters, and the whole of that of her sons."

We may here interpolate the following comments:—

(1) During a period of about 40 years prior to the impositions of sanctions on Italy in 1936 Italian queens were freely imported to Great Britain and a large proportion of the native bees became hybridised. British queens however remained in the majority at any particular time, their cross-mated daughters producing black drones. The consequent parthenogenetic tendency to reversion to the pure races, supported by environmental conditions, persisted, and was reinforced by the sudden cessation of Italian importations. Consequently we now see a rapid reversion to bees of the British type and a corresponding diminution of the Italian bees and their hybrids.

(2) When we select a breeding queen in one year we are at the same time selecting drones of the following year, and as these drones are likely to mate with young queens in apiaries other than our own, the effect on neighbouring bees generally is involved in the selection we may make.

Drone-breeding stock.

Drone-breeding is easier than queen-rearing but hardly less interesting. Considerable knowledge and attention are needed to ensure that a supply of choice drones is available throughout the breeding season, and that none but these have much chance of mating with the young queens. The serious queen-breeder therefore does not leave drone production to nature. He raises large numbers from his best stock, maintains the supply until autumn, restricts drone breeding in his other stocks, and does his best to ensure that only drones of the selected strain pervade the neighbourhood of his queen-rearing apiary. We may now consider what steps he takes.

To obtain and maintain a plentiful supply of select drones.

About the middle of April, weather permitting, choose a prosperous stock, the queen of which has the requisite good qualities, and after removing the least desirable comb, insert into the brood-nest a wired frame fitted with a starter (narrow strip of foundation), or a drawn-out drone comb. The starter will be developed into a full comb with drone-sized cells, and if the stock is fed, will be quickly filled with drone brood. A second starter or drone comb can then be inserted, and if both are maintained in the stock throughout the season a large population of drones will constantly be present. The difficulty of swarming will be unlikely to arise on account of the presence of the large drone population (pp. 206–207).

Periodical removal from the hive of a comb of sealed drone brood together with one or two combs of matured worker brood, and the substitution of corresponding foundation for them will prevent the stock from becoming strong and will of course make it useless for honey storing. The removed brood should go to the building up of a queenless stock to be kept alongside the parent stock in the queen-mating apiary. The queenless stock, which for convenience we may call a drone-bearing stock, will keep its drones as long as it remains queenless.

Drone-bearing stock.

To maintain a stock in a queenless state however is not altogether easy. If it is allowed to rear a queen from young brood inadvertently given to it, the drone population will not be long maintained after she begins to lay. If it is not able to rear a queen it will, sooner or later, develop laying workers (p. 152). As it may be needed in a queenless state for three or four months it may be allowed to rear a succession of young queens—about one a month—which must be removed when they begin to lay. Periodical addition of drone and sealed worker brood, together with sufficient feeding, will maintain the strength of the stock, and its drones will be retained until it is allowed a permanent queen at the end of the season. It may here be noted that a drone-bearing stock must not be strengthened with much unsealed brood. Its larval food, needed to support the drones, would be diverted to the young brood, and the drones would diminish or disappear.

It is not easy to induce a fertile queen to lay drone eggs after August. She may do so if liberally fed, but if the weather becomes unfavourable the bees will not allow the eggs to hatch and will remove them.

In the course of queen-rearing we find that a proportion of the young queens fail to become mated. If these are not removed they will ultimately begin to lay eggs which produce only drones. These will be bred in worker cells and consequently will be undersized and comparatively weak. They are possibly capable of fulfilling the marital function (p. 92), but being small and produced in limited numbers are unlikely to be successful in competition with the numerous drones of a drone-breeding stock.

An unmated queen is a failure and is usually destroyed by the beekeeper when discovered, but the drone raiser may have a use for her. Introduced to the drone-carrying stock referred to in the preceding paragraphs, she will keep it contented, prevent it from developing laying workers, and will enable the bee-keeper to maintain a large force of drones as long as he can induce his selected stock to produce the necessary drone brood. Her own drone brood however will appear mostly if not entirely in worker cells and if allowed to emerge will give rise to undersized drones. It should therefore be removed before maturity. An inspection of each drone-bearing stock should be made every three weeks.

Drones will pass through the winter with an unfertilised queen if the stock has been provided with sufficient food and worker brood. In the following spring the stock may again be provided with sealed worker brood and used for carrying selected drones.

If no drone-breeding queen is available it is easy to secure one. All that is necessary is to prevent one of the earliest batch of young queens from becoming mated by placing a piece of excluder over the entrance of the nucleus hive for a month. Feeding will encourage her to lay and her unmated condition will be confirmed by the raised cappings of her first sealed brood. This brood must be removed and her little stock can then be built up into a drone-carrying stock by adding to it from time to time mature drone and worker brood from the stock of the selected drone-breeding queen.

The drones from these two stocks will be sufficient for the needs of a fairly large queen-rearing apiary.

The undersized drones raised in worker cells from queens which have failed to mate, and those reared from the eggs of laying workers, should not be kept in a queen-mating apiary. Structurally complete and having spermatozoa which are indistinguishable from those of normal drones they have been shown to be fertile by artificial insemination (93, p. *155*). They are possibly able to mate with queens, but whether effectively or not is doubtful. In any case it is obviously undesirable that they should have the opportunity.

It must always be remembered that a stock raising

or carrying large numbers of drones needs much food and should not be expected to gather surplus honey. The duties of its nurse bees are exhausting for they not only have to feed the normal brood but probably also have to provide some brood-food for maturing drones (p. 87). Feeding should therefore be more or less continuous outside the period of nectar-flow so that the drones may be amply nourished during their early development and always lusty and vigorous during their adult life.

A queen whose store of spermatozoa is almost depleted shows signs of failing by laying an increasing proportion of unfertilised eggs which produce drones. If she is of approved race and strain she may be used to head a drone-carrying stock.

Restriction of undesirable drones.

A queen may mate with a drone from her own hive or from any other within a distance of two or three miles. She is most likely to mate with one from a hive in the near neighbourhood, hence the importance of a queen-mating apiary which is reasonably well isolated from other bees (pp. 303–305).

If however we must rear our queens without isolation we must endeavour to prevent the production of undesirable drones. The best way to do this is to requeen all the hives in the apiary, together with those of our near neighbours if possible, with young queens of an approved strain. In the following year all the drones in the neighbourhood will have qualities of this strain and it will be necessary to restrict only those in the stocks which prove to be least satisfactory.

As beekeepers are reputed to be amongst the most friendly of people it should be easily possible to run local co-operative schemes for systematic requeening and improvement of stock. Without co-operation the individual beekeeper who rears his queens at home is constantly handicapped.

It is in the economic interest of the beekeeper, apart from considerations of queen-rearing, to limit the production of drones. This is done by removing from the hives, in spring or autumn, all combs containing more than a few drone cells, and substituting for them sheets of worker foundation. On these the bees build worker cells only, but as the combs become older some of the worker cells, particularly those at the bottom and lower corners of the comb, are broken down to make room for drone cells which the bees are instinctively urged to provide. If the combs are not periodically renewed for a few years a hive will contain a large proportion of drone cells, and therefore a large drone population every summer.

According to Cheshire (17, II, *106*) a comb containing sealed worker brood on both sides has a thickness of approximately $\frac{15}{16}$ of an inch. The minimum bee space between one comb and the next is $\frac{5}{32}$ of an inch. The total minimum lateral space occupied by a comb and a bee-space on each side is therefore $\frac{15}{16}'' + \frac{5}{32}'' + \frac{5}{32}'' = 1\frac{1}{4}''$ and this is the distance between the midribs of adjacent combs, there being a double bee-space between them.

Drone comb with capped brood on both sides is $1\frac{3}{10}''$ thick which is more than $1\frac{1}{4}''$. In natural comb building bees build their worker combs from $1\frac{9}{20}''$ to $1\frac{1}{2}''$ apart. This allows them to build some drone comb, but only by encroaching on the passage-ways between neighbouring combs.

The familiar W.B.C. "metal ends" separate combs by $1\frac{9}{20}$ or $1\frac{1}{2}$ inches when placed end to end, but only by $1\frac{1}{4}$ inches if alternately overlapping. When foundation "starters" are given to a swarm therefore the frames should at first be spaced $1\frac{1}{4}$ inches apart. The bees will then be constrained to build entire worker combs, not finding space for drone cel's. When the combs are completed the beekeeper separates them to the natural distance of $1\frac{1}{2}''$ apart.

As most beekeepers now use full sheets of foundation

there is no need of the narrow spacing to restrict the building of drone comb.

Except in early spring the bees of a normal stock will fill vacant spaces or repair damaged combs with drone cells. In practice it is impossible to ensure that a stock will produce no drones during a whole season. We can only keep them down to very small numbers.

It is sometimes stated that a stock with undesirable qualities may be deprived of its drones in the following way:—

Place an excluder guard over the hive entrance. Cage the queen temporarily. Remove all the combs and shake the bees on to a large alighting board in front of the hive. The workers enter through the excluder but the drones cannot do so. Replace the combs as shaken and finally restore the queen. The excluded drones cluster at the entrance and may be disposed of the next morning.

This rather distasteful procedure is not as simple as it seems. The drones choke the excluder and a considerable time may elapse before all the workers have regained the hive. Numbers of the disturbed and excited drones will manage to re-enter the hive each time this is opened to receive a shaken comb. Combs containing drone brood must be removed altogether and worker combs or foundation substituted for them, or a fresh crop of drones will be bred.

Drone traps are sometimes used:—

(*a*) to capture unwanted drones.
(*b*) to catch the drones from a selected stock for transport to the queen-mating apiary.

The essential features of a typical drone trap are as follow:—

A light box of the same length as the hive entrance, against which it is fitted, is divided into two chambers, the upper one larger than the lower.

The wooden floor between them is fitted with two or three cone escapes projecting upwards into the top chamber. This chamber has a front panel of excluder zinc, and a movable wooden cover.

The lower chamber is divided into two parts by a sheet of excluder zinc which slopes from the top of the front to the bottom of the back of the chamber. This chamber has no front panel (Pl. XII).

Worker bees can pass freely out of the hive, either

(1) directly through the sloping sheet of excluder, or
(2) along its upper surface, by way of the cone escapes into the upper chamber, and out through the excluder panel. They return via (1).

The drones cannot take course (1) and are therefore obliged to take course (2). Arrived in the upper chamber there is neither escape nor return for them and they are trapped.

Pratt (66, pp. *32* and *40*) describes a simpler drone trap which works quite well. A framed queen-excluder is placed under the brood-box so as to allow the worker bees to pass through it and in and out at the usual hive entrance. For the passage of the drones a 1″ hole is bored through the front wall of the hive, about an inch above the entrance. Over this hole the trap is placed. This consists of a long shallow box of approximate dimensions $17'' \times 4'' \times 4''$. The front is of excluder zinc and the back, which fits against the hive front, is a detachable wooden panel. A hole is bored in this panel, to cover exactly the hole in the hive wall, and on its inside surface is fixed a cone escape pointing into the trap. Drones pass from the hive, through the cone, and into the trap whence they can neither escape nor return. The trap is to be securely fixed to the hive for a whole day on which bees are flying well, and removed at evening.

PLATE XII

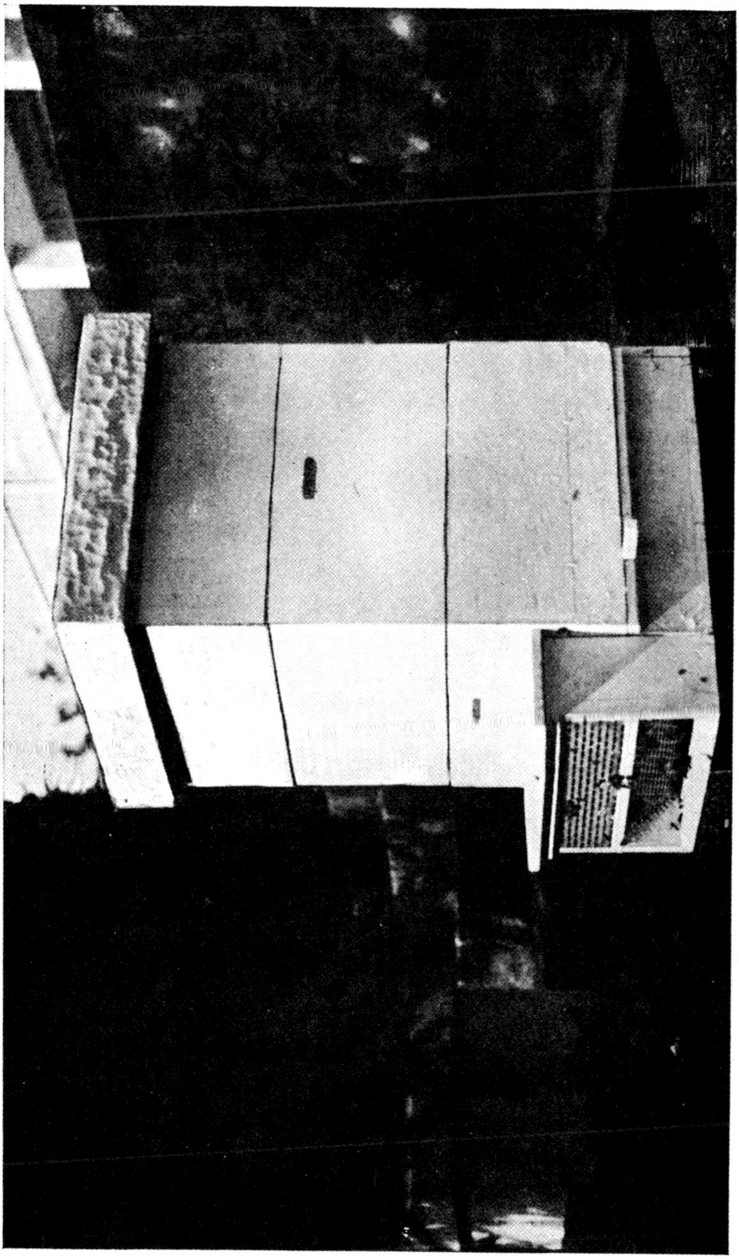

[*Photo: L.E.S.*

DRONE TRAP FIXED TO ABSTRACT DRONES FROM A STOCK TO BE USED FOR COMPLETING QUEEN-CELLS

Trapped drones should not be imprisoned longer than is absolutely necessary as they soon worry themselves to death. If destined for the queen-mating apiary they should be trapped during the whole of a warm afternoon and introduced to a drone-carrying stock during the same evening. They quickly starve when food is withheld from them and it is therefore necessary to sprinkle them with sugar syrup through the excluder before they are taken on a journey. They may be introduced to their new home by removing the lid of the trap and inverting the latter over the feed hole, or by shaking them directly into the brood nest. They are welcomed in any queenless stock or nucleus, and also in a queen-right stock except when drones are being excluded.

The presence of drones in a stock is inimical to queen-rearing because they are consumers of the brood-food secreted by the nurse-bees (p. 206). In the illustration in Pl. XII the drone trap is abstracting drones from a stock to be used for cell-completing.

In all places where queens are being raised the drones should be suppressed in stocks that are vicious, diseased, or that produce abnormal forms,—e.g., albinos (9, p. 59), and the undersized progeny of unfertilised queens and laying workers should be removed from the hives whilst in the sealed brood stage.

CHAPTER VI

VARIETIES OF HONEY BEES

A BREEDER of queens who desires to improve and maintain the quality of his stock by selection must decide in the first place what variety or strain he should cultivate. He should be acquainted with the general merits and de-merits of the races which have been found to be of commercial value. He should also take into consideration the suitability of his district for one race rather than another, as well as the general type of bees kept by others in his neighbourhood. If, for example, most of his neighbours keep black bees he cannot hope to breed and maintain pure Italian stock. In Chapter VII it will be shown that it is next to impossible to perpetuate the characteristics of a particular hybrid stock even with assured isolation, and that a breeder should therefore attempt to cultivate bees of one of the pure races, within which variation is limited.

Of the genus *Apis* (bee) there are several species, one of which is *Apis mellifica* (the honey-making bee) sometimes referred to as *Apis mellifera* (the honey-bearing bee). The former title seems to be more appropriate since the bee makes honey from nectar. This species comprises a considerable number of varieties, each of which, on account of mutations (p. 130) favoured by environment, in the course of ages has acquired certain fixed characteristics which distinguish it from the others. These varieties are often referred to as races of the honey bee.

It will be sufficient here if we consider the general characteristics of the varieties which have proved to be of commercial importance in temperate climates, viz., the common Brown or German bee (including

British "blacks"), Italians, Caucasians, Carniolans, and Cyprians. Amongst the less known varieties, —not kept in this country—are Egyptians, Syrians, Palestinians, Saharans, Tunisians, and Chinese bees. There are also the bees of India, including *Apis indica*, *Apis dorsata*, and *Apis florea*, which are not of the species *Apis mellifica*. For descriptions of all these the reader is referred to Herrod-Hempsall (33, II, p. *1517* et seq.) or Root (72, p. *604* et seq.).

Of the common black bees and Italians, together with their hybrids, the writer can claim wide experience. During forty years of beekeeping he has attempted also to use Carniolans, Caucasians, and Cyprians. The Carniolans were given up because of excessive swarming. The so-called Caucasians turned out to be hybrids and hardly different from British-Italian hybrids. The one queen obtained from Cyprus bred throughout a mild winter so that her previously well-provided stock starved to death—in the presence of a large brood nest—in February. He therefore cannot speak from adequate experience of the three last-named races, and the descriptions of them which follow are partly based on a comparison of the views expressed by other writers.

Before considering these varieties in detail we may premise:—

(1) That each variety has in the course of long ages become to some extent adapted to the climate, flora, and other conditions of the regions in which it is indigenous, and that consequently it may not be equally prosperous when transported to a totally different environment. For example, some of the Mediterranean races, accustomed to a warm climate and a nectar-gathering season which lasts for a great part of the year, are prodigal in brood-rearing and food consumption, and when transferred to

colder climes often fail not only to yield surplus honey but also to survive a long winter after an indifferent season. The native bees, with a shorter breeding season, economise their stores, go earlier into their winter rest, and usually have a reserve of food with which to face the critical spring months.

(2) That within each variety there are sub-varieties distinguished by differences in behaviour—e.g., British and Dutch "blacks".

(3) That what may be called minor variations give us different strains within each variety. Some strains are more industrious, more amenable, less inclined to swarm, or are more productive than others. It is with this question of "strain" that the queen breeder who would improve his stock is specially concerned. That the breeder of queens of a single race is not helped by Natural Selection over a relatively short period is illustrated by the following:—(57, p. *107*).

About sixty years ago several colonies of Italian bees, not specially selected except for purity of race, were taken to Kangaroo Island, South Australia, for the purpose of starting a breeding scheme, the island not being inhabited by any other bees. The scheme had to be abandoned, but the bees were left in the island.

They multiplied, and numerous colonies now live in the hollow trees of virgin forest areas. An attempt is now being made by the South Australian Department of Agriculture to establish a "stud farm" where by artificial selection improved breeding stock may be raised from these bees.

Ophel's account of them shows that sixty years of wild life has not resulted in any visible tendency to uniformity in their characteristics but that the differences which ordinarily distinguish strains of Italian bees still persist. Of them he says:—

"The variations are not all apparent at first sight; some are more thrifty than others—"better doers" in common parlance; some require a puff or two of smoke when opened, while others need none; some build burr comb or propolise more freely; in some the drones show their bronze lines on the edge of abdominal segments whereas in general the drones are evenly black on top with light bronze underside. Observations so far indicate that, in colonies with the former type of drone, the workers are not quite evenly marked, and are inclined to be restless on the combs."

(4) That it is extremely difficult to keep imported bees of a foreign race pure for any length of time. Sooner or later the queens become mated with native drones and although the resulting hybrids may sometimes show improved stamina, they often especially in the second and later generations—show deterioration and become distinctly unpleasant to manage.

(5) That a description of any variety of bees can be general only, and subject to exceptions.

The common brown bees.

The brown (sometimes called "black") bees are indigenous to almost the whole of Europe. In the 17th century and later they were taken to North America and to what are now British dominions. When considering their characteristics we must remember that they comprise strains which show marked divergencies in disposition and economy, varying from the excellent bees of which the British natives are typical to the almost worthless heath bees of the Luneberg district of Germany, some strains of Dutch bees, and the unpopular brown bees of North America. One description therefore will not apply to them all.

Until the importation of Italian bees, which began in 1859, the British Isles were occupied by only one variety of hive bees—now commonly referred to as British "blacks". They were everywhere plentiful, had become adapted to our variable climate, and had many excellent qualities. The desire to possess the brighter coloured and more prolific Italians became wide-spread, and for more than fifty years they were imported in increasing numbers with the result that most of the bees in the country were hybrids. About 1904 the "Isle of Wight" Disease epidemic appeared and by 1917 it had destroyed a large proportion of the bees, British, Italians, and hybrids alike. An extensive scheme of restocking was carried out in 1917–18 by the Board of Agriculture, Dutch bees being imported for the making of nuclei, and Italian queens to be introduced to them. Great numbers of Italian queens continued to be imported at low prices and such was the extent of hybridisation that many writers have expressed the view that bees of pure British origin cannot now be found. The writer does not hold this view. Apart from the fact that he has continuously found British bees in certain country districts showing no sign of crossing with foreign races, the laws of heredity (Chapter VII) conflict with the supposition that a pure race can be eliminated by crossing alone. In 1936 sanctions were imposed on Italy by the British Government and the importation of queens from that country diminished from that time and ceased during the war. For some years, too, the importation of other races, Carniolans, Caucasians, etc., has been discontinued. The Italian element, as shown by colouring, is steadily disappearing and many of our bees are becoming dark and indistinguishable from the old British bees.

Incidentally we may say that some people have associated the cessation of imports of queens from the continent with the striking diminution of Nosema disease which some years ago was widespread in the British Isles. In a Rothamstead Research Station report on two surveys of Adult Bee Diseases involving the examination of about 1,100 samples of bees collected at random from all parts of England and Wales during the years 1941–2 and 1943–4 Dr. Butler showed that Nosema had been found in only eight of the samples (15, p. *348*). The disease however was common to both native and imported bees, and it is quite likely that its diminution has been due to natural causes.

The common brown bees, of which the British bees are an excellent sub-variety, are distinguished by the dark brown—almost black—colour of the dorsal portions of the thorax and abdominal segments. The latter are bordered by bands of greyish brown hairs which give the whole abdomen a brownish appearance. When these hair bands have disappeared—as they sometimes do on account of age or a depilatory disorder (17, II, p. *568*) the abdomen looks black.

The British "blacks" are hardy, acclimatised, and winter well. They are industrious, working early and late on favourable days. They are moderate in their tendency to swarm and compare favourably with the yellow races in this respect.

For the production of comb honey they are unrivalled, taking readily to sections and sealing their combs with beautiful white capping. They use little propolis. They are comparatively resistant to disease and as Maisonneuve says, "reserve their stings for the awkward". During manipulations they hold well to their combs and as a rule do not run about excitedly whilst being examined.

They do not waste their substance in unnecessary breeding, as do some other races, for they are rather slow in developing in early spring, do not develop excessively large brood nests, and settle down for winter early in the autumn. Consequently their brood nests, diminishing as autumn proceeds, leave room for the accumulation of winter stores. They consume little food during the winter and usually reach the spring time with some reserves.

On the other hand they are inclined to pillage whilst being but indifferent defenders of their own homes, and, in contrast with the Italians, are comparatively tolerant of the wax moths.

They are good honey gatherers and will store surplus even in bad seasons. Some of the softer and more prolific foreign races are highly productive in good seasons but fail in bad ones, and since a majority of our bee seasons are poor the British bee is, in this respect, to be preferred in our variable climate. Of it Herrod Hempsall (33, p. *1522*) says, "For nectar-gathering it excels any other race . . . the sooner it is established as the dominant race, the better it will be for apiculture in Great Britain"; and Root, (72, p. *609*) speaking of the American preference for Italian bees, says "whether they will ever displace the common brown bees that have done so well in those countries (European) is doubtful".

Average British blacks are less patient of manipulation than Carniolans, Caucasians, and pure Italians, but it is possible, by selection and isolation, to evolve British strains which are as docile and productive as the best of other races.

Certain strains of brown bees, particularly those imported from some parts of France, and occasionally from Holland, show extreme nervousness when disturbed and exposed to light. They run rapidly over the combs in all directions, collect in lumps, and drop to the ground as though in terror. With such bees, manipulations, especially the finding of the queen, are difficult. Root (73, p. *554*) considers that the black bees of America are of this type and that they are of Dutch origin. It is little wonder therefore that our beekeeping friends in the United States prefer their American-bred Italians.

It is generally agreed that there were no honey bees in America before the advent of the European settlers. The Red Indians called them "the white man's fly" and anticipated the white settlers' advance through the country by the previous appearance of their swarms of bees.

"Wheresoe'er they move, before them
Swarms the stinging fly, the Ahmo,
Swarms the bee the honey-maker;
Wheresoe'er they tread, beneath them
Springs a flower unknown among us,
Springs the White-man's Foot in blossom."
(*Longfellow*, *Hiawatha*, *XXI*, *197–202*.)

Since white clover, the best of honey-yielding plants, was introduced to America by the early settlers (34, p. *130*) one is naturally inclined to suppose that Longfellow might be referring to its blossom in this passage, but Pellett (61, p. *2*) citing Barton (1793) calls attention to the fact that John Elliot, who translated the Bible into a language of the Indians, noted that they called the broad-leaved plantain "the Englishman's foot" because it sprang up wherever the Englishman went. Langstroth (45, p. *289*) however accepts the view that the poet referred to the white clover.

The Pilgrim Fathers had assembled in Holland before they sailed from Plymouth in the "Mayflower" in 1620. Amongst other domesticated animals they probably took bees with them, perhaps from Holland. Other colonists included French and Dutch emigrants who also would be likely to take bees with them from their mother countries. Pellett (61, p. *3*) gives an interesting account of the early records of the presence of honey bees in America, the earliest of which is dated

1640. Nothing definite appears to be known of their earliest arrival.

Whilst the Americans therefore have good reasons to prefer Italians to "blacks" British beekeepers have not the same reasons for disliking the latter, for native British bees do not usually exhibit the above-mentioned nervousness nor are they addicted to excessive swarming as are the Dutch and German heath bees.

Italian bees.

Of the "yellow" varieties of bees inhabiting countries to the North and South of the Mediterranean the Italians are of chief commercial importance. Spinola, who described them as a "new and rare species", found them in Liguria, and therefore named them *Apis Ligustica* (45, p. *293*).

Some writers have surmised that they are the result of crossing yellow bees, brought by the Romans from Egypt or Cyprus, with the common brown bees, and that their characteristic colouring has become fixed through isolation imposed by the Alps and the Mediterranean sea. This however is most unlikely for in such a case the hybrids would tend to revert to the pure types (p. 147). We know too, that the Roman writers Vergil (89, ll. *95–99*), Varro (88, III, *16*), and Columella (18, IX, 2), were familiar with the golden as well as the dark bees, and we may therefore presume that the bees of Italy two thousand years ago were similar to those of to-day.

Whilst it is likely that they are related to other Mediterranean yellow races their distinguishing characteristics are probably the results of mutations (p. 126) which happened in more remote ages.

Two types of Italian bees are recognised, the yellow or golden, and the tawny or leather-coloured. The former were first introduced to the United States in 1858 and to England in 1859. Regular importations to both countries followed, the Americans finding the

Italian bees superior in docility and working qualities to their unsatisfactory "blacks", and the English preferring them largely on account of their prolificacy and attractive colouring. They have now spread to most parts of the world and in countries where seasons are reliable and honey yields high, and where, as in the United States, they have been improved by artificial selection and acclimatisation, they are not excelled by bees of any other race.

The typical yellow Italian is distinguished by yellow bands on the anterior portions of the dorsal plates of the first three abdominal rings or segments. These plates are crossed by bands of light yellowish hairs and their lower or posterior edges are black. As in queen-rearing it may be important for us to identify pure Italians and to distinguish them from their hybrids the following detailed description of their characteristic marking—given by Cheshire (17, II, p. *611*) may be helpful:—

"The first abdominal ring on the dorsal side mainly faces the thorax and may be missed by careless observation; its lower edge only is black. The upper two-thirds of the second is yellow, the upper third smooth and hairless because this passes beneath the ring when the body is contracted. A band of yellow hair covers the second third and adds much to the beauty of the bee as hairs and ground are alike yellow. The lower third of the ring is glossy black, carrying many microscopic hairs and a minute fringe. The third ring resembles the second, while the fourth and fifth carry yellowish hairs, but are black. The sixth ring, black also, is nearly hairless." The under side of the abdomen is light brown. The small posterior portion of the dorsal covering of the thorax,—the scutellum—is of a dull orange colour.

The queens vary much in colouring. Some show much yellow—others little. By persistent selection for colour, queens have been produced both in America and

England showing yellow bands on five of the six abdominal rings, the sixth (terminal) one being black. Such queens and their progeny have been boosted as "Golden Italians" but they have not shown any particular merit apart from their beauty.

Typical Italians, imported direct from Italy, have the following characteristics:—

They are gentle, easily subdued, and adhere to their combs during manipulations. They are therefore easily managed. They are active, working early and late, and, according to some authorities, fly to greater distances than the brown bees in search of their food. In favourable conditions of weather and district they store large crops of honey. They begin to breed early in the spring and as their queens are very prolific they develop large populations which they maintain by continuous breeding until late in the year. They are rather more inclined to swarming and raise more queen-cells than British "blacks". They do not tolerate the wax moth in their hives and whilst they defend their own homes well they are themselves efficient robbers.

Amongst their defects we must notice the following:—

In poor seasons (and we in Britain have many) they consume most of what they gather in excessive brood rearing. Especially is this the case during the spring and autumn. During a mild October, for example, they will often convert their winter stores into brood, and unless heavily fed, may perish in the ensuing winter. In general they need more feeding than the thrifty dark bees.

As a rule they are less suitable than the dark races of bees for the production of comb honey. They are less inclined to occupy and work in sections, and they seal their honey with cappings which being directly in contact with the liquid honey beneath them, often lack the attractive whiteness of the sealed combs of the dark bees. In some seasons however their comb honey sealing may be perfect.

When crossed with dark bees—as sooner or later becomes inevitable—their hybrids are usually good in the first generation but in later ones are often aggressive and even vicious. It is supposed by some people that disposition is derived from the drone parent and that if an Italian drone mates with a dark queen the progeny are amiable but that if a dark drone mates with an Italian queen their bees are likely to be ill-tempered. There is however no scientific support for this theory.

The darker ("leather-coloured") variety of Italians is reputed to be rather less docile, more hardy, and more suitable for cool climates than the yellow variety.

The purity of a stock of Italian bees is judged by the markings of the workers. If the great majority of these show three yellow abdominal rings the queen is considered to be pure, and purely mated. If only one or two yellow bands appear the bees are deemed to be hybrids. Similarly if a proportion of the bees of a dark stock show one or more yellow bands the stock as a whole is considered to be hybrid.

The bands can best be examined by causing a well-fed bee to walk on the inner surface of a window. The rings of the distended abdomen are then further apart and the colours are easily distinguished by the transmitted light.

Carniolan bees.

The Carniolans are a dark race of bees inhabiting a large district bounded on the West by the Carnic and Julian Alps and on the North by the Noric Alps. These mountain barriers have tended to isolate them from the Italian and German bees and to promote the development of characteristics entitling them to be considered a distinct variety.

They are rather larger than the common brown bee and as dark in colour except that the four bands of hair crossing the dorsal abdominal plates are thick, silvery-grey, and so prominent that they brighten the appearance of the bee and make the separation of the abdominal rings more apparent.

They are energetic workers, make beautiful white-capped combs, use little propolis, and winter extremely well on comparatively little stores. They build up rapidly as the swarming season approaches. They are perhaps the gentlest of all bees and can be handled at all times without smoke or other intimidation.

Unfortunately they have one characteristic which has prevented their wide adoption. They are given to excessive swarming. Their swarms, and even their second and later swarms will build up rapidly and swarm again in the same season. This undesirable tendency can be moderated by selective breeding, large hives, and shade, but these bees are so uncertain and difficult to keep at home that serious beekeepers in this country will not be bothered with them.

Caucasian bees.

Caucasians are of two kinds a dark race predominating in the mountainous regions and a yellow one mainly found in the lowlands of Caucasia. Queens of the former have been imported to western countries where they have attained some degree of popularity. Direct importations to western Europe however are difficult on account of distance and it is doubtful whether, owing to political circumstances, any have been made for many years. Those that have reached Great Britain have mostly come from nearer countries, and in some cases have been hybridised by brown bees on the way.

Their workers are slightly smaller than those of the brown bees but are almost indistinguishable from them in colour. The light pubescence on their abdominal tergites is prominent, and gives these a greyish appearance.

They are industrious, hardy, winter well, and are

extremely gentle. Some say they are the gentlest of all bees. They are little inclined to robbing and except for a marked tendency to swarm, are easily managed. Their great fault is that they collect and use large quantities of propolis. As Root (71, p. *610*) remarks, "They deposit a large amount of propolis, sticking large chunks of gum in all parts of the hive". This, to say the least, is very inconvenient to a manipulator, especially in Autumn. Abbott, (1, p. *84*) however, reporting favourably on his experience of these bees in England, states that apart from depositing propolis near to the hive entrance they did not use it to excess in other parts of the hive.

These bees are favoured largely on account of their docility. Maisonneuve (62, p. *49*) considers that as honey-gatherers they are not equal to the Italians, Cyprians, or French brown bees.

Cyprian bees.

Owing to their complete isolation the bees of Cyprus are regarded by some authorities as the purest of the yellow races, and possibly the mother-race of others, —including Italians. Rather smaller and more slender than the Italians they have similar markings, but the first three abdominal rings are orange to bronze coloured, the pubescence russet brown, and the under-side of the abdomen pale yellow. The scutellum is of bright orange colour and conspicuous. Their drones are smaller than those of the brown or Italian bees and are brightly coloured. They are considered by some to be the handsomest of the varieties of the hive bee. In their own country they have to work hard to obtain a living and are consequently exceedingly energetic. Introduced to more favourable countries they are great honey-gatherers and according to Benton (7, p. *15*) they hold the record for having produced the greatest weight of honey from a single stock in one season. Their queens are very prolific and they winter well if provided with sufficient food. They raise many queen cells when preparing for swarming. Their comb honey, like that of some Italians, is not of a desirable appearance. They have one unfortunate characteristic which has led to their general rejection outside their own country. They are extremely irritable when manipulated and especially when smoke or other intimidants are used. Root describes them as "awfully vicious" (73, p. *559*) and Cheshire (17, II, p. *615*) writes of them,—"It is true that, when thoroughly irritated, they make war with a will, and the bee-master needs the philosophy of a stoic, or the hide of a rhinoceros, if he is to stand his ground, for the enemy will submit under no treatment that does not involve decimation". Cheshire however goes on to say that he found they could be handled if approached gently and without intimidants. Many years ago the writer's father introduced an imported Cyprian queen to his apiary of about eighty stocks of British "blacks". Her stock was not unmanageable but in two or three years the resulting hybrids became a menace to the neighbourhood when manipulations were necessary.

Choice of a Variety.

It is hoped that the foregoing descriptions of the chief foreign races of bees are sufficiently full and accurate to enable a beekeeper to make a wise choice between any one of them and the native brown bee. He should especially take into account our climate, the abundance or scarcity of the bee-flora of his district, the problems of management, the kind of bees preferred by experienced local beekeepers, and the possibilities of local co-operation in breeding. If there is a local effort to develop any special race or strain he should pause before he decides to impair this effort by the introduction of queens of another variety.

It is doubtful, all things being considered, whether any advantage is gained by the importation of foreign races. Their defects may outweigh their merits and their hybrid descendants may be unpleasant to manage. It is therefore good policy to concentrate on the development of a good strain of local (native) bees. If foreign bees must be introduced to a locality they should not be imported from a tropical or sub-tropical country but should have been bred through several generations in countries which experience winter conditions at least as severe as our own. It is generally known for example that Italians raised in this country or in the U.S.A. are on the whole superior to those directly imported from Italy. Indeed Phillips (63, p. *8*), writing of the Italian bees as long ago as 1905, remarks that "very superior stock is reared in the United States, and queen bees of the Italian variety are actually shipped from this country to Italy to be used as breeders".

CHAPTER VII

HEREDITY IN BEES

THE queen-rearer who seeks improvement of his stock by artificial selection or who desires to maintain the best qualities of an approved strain cannot do better than to make some acquaintance with the mechanisms by which physical characteristics are transmitted, and the laws of heredity which arise from them. These will not only be of some practical value—although less than in the case of other forms of animal stock where mating is easily controlled—but they will also greatly increase his interest in animal and plant life generally. Particularly will they serve to show the limitations under which the breeder of queens must work, and why it is difficult to maintain a pure race of bees or to establish and perpetuate a strain of hybrids with fixed characteristics.

In this connection we may premise that the beekeeper is handicapped by the impossibility of strictly controlling natural mating, by complications due to plural mating, and by the fact that, owing to the parthenogenetic production of the drone the possibilities of line-breeding are limited since a queen has no full brother and cannot be mated to her father or son.

Within the limited scope of this work consideration of such a vast and involved subject as heredity must necessarily be brief and elementary. We shall therefore confine ourselves to essential elementary principles. Readers who desire to study these matters more deeply should consult modern works on Biology.

Body cells.

In Chapter III we have referred to the cells of which plant and animal tissues are composed, and in particular have considered the ovum and spermatozoon as specialised cells which by their union give rise to a new living organism.

In the lowest forms of life there is no union between male and female cells and new individuals arise from the division of single cells. In the higher forms growth and renewal of tissue are maintained by similar cell-division, but reproduction is effected by the union of cells from separate individuals. The apparent value of this will be considered later.

Chromosomes.

Every cell in a living body, except the red corpuscles of the blood of most mammals, contains within its mass of protoplasm a smaller specialised mass called the nucleus. This contains a number of minute bodies called chromosomes. These in some stages can readily be stained and made visible. It is within these chromosomes that the unit elements or hereditary factors, which give the various parts of the whole body distinguishing characteristics, are located.

The number of chromosomes in a cell of a particular species of any animal or plant is generally constant. They usually comprise a double set, one set being derived from the male and the other from the female parent. For example the number in a human cell is 48 or 24 pairs; in a female honey-bee, 32 or 16 pairs; in the vinegar-fly (*Drosophila melanogaster*) 8 or 4 pairs,—(it is from the study of this fly that much of the knowledge about chromosomes has been gained); in the fuchsia 22 or 11 pairs; in the evening primrose 14 or 7 pairs; and so on. There are however important exceptions to this general statement.

The development of a new individual normally needs the union of male and female gametes (p. 123) and consequently the presence of the full number of chromosomes characteristic of the body-cells of the particular plant or animal. Nature however provides a few examples of the complete development of individuals from unfertilised eggs, which contain only half the normal number of chromosomes. Such individuals are described as haploid, whilst those derived from the normal or double set of chromosomes are called diploid. Thus the queen and worker bee are diploid with body-cells each containing 32 chromosomes whilst the drone is haploid with cells containing 16 chromosomes.

Genes.

Although the chromosomes themselves are extremely small each carries on it, in a linear series comparable to a string of beads, a large number of unit elements called genes. These are sub-microsopic—yet their existence, and in some cases their relative positions on a chromosome, have been ascertained. Each is responsible for some physical characteristic in the organism of which the cell forms a part. Normally a gene is unchangeable, —its characteristic persists from generation to generation notwithstanding its apparent temporary disappearances in what is known as "recession". (p. 139).

Sex chromosomes.

The sex cells of animals and plants are technically known as gametes. The result of the union of the nuclei of male and female gametes is called a segmentation nucleus, and the cell of which this forms a part is known as a zygote. This develops into a male or female embryo according to its chromosome constitution.

In the gametes of most animals and plants many of the chromosomes carry genes responsible for sexual characteristics but there are two chromosomes which are sex-determining. These are usually distinguished as accessory or X and Y chromosomes.

The female parent forms X bearing gametes (eggs). The male parent forms both X bearing and Y bearing gametes (spermatozoa) in equal numbers.

If an egg is fertilised by an X bearing sperm the

resulting zygote carries XX chromosomes and develops into a female embryo. If the egg is fertilised by a Y bearing sperm the zygote carries XY chromosomes and becomes a male. Since X and Y sperms are produced in equal numbers by the male parent, notwithstanding that most eggs and sperms are wasted the aggregates of males and females produced are approximately equal.

There are many exceptions to this general rule. In birds, butterflies, and moths, for instance, it is the female parent which produces the two kinds of gametes. In bees, wasps, and ants, no Y chromosomes are formed. All the sperms and all the eggs are X bearing (p. 158).

Parthenogenesis.

In 1845 Dzierzon propounded the theory of parthenogenesis (birth by a virgin) as a result of his observations relating to the origin of drones. He had observed that the eggs of unmated queens and of laying workers (incapable of being mated) produced drones, and concluded that the eggs of the queen produced drones if they were not fertilised by sperm from the spermatheca. On the assumption that sex chromosomes are involved (p. 158) a cytological explanation of this phenomenon is briefly as follows:

The 16 chromosomes of each matured egg-cell (p. 128) include one X chromosome. Each sperm-cell also contains an X chromosome. Every *fertilised* egg therefore contains (15+1X) + (15 + 1X), or 30 + 2X chromosomes, 2X being the essential condition for femaleness. Every unfertilised egg contains 15 + 1X chromosomes, but this, lacking the XX chromosome factors necessary for femaleness, becomes a male.

The drone therefore has no father and inherits only characters derived from his grandparents through his mother.

The development of haploid males from unfertilised eggs is common in the Hymenoptera which include bees, wasps, and ants.

Allelomorphism.

We have already spoken of two sets of chromosomes in a cell of a living organism, and in particular of the two sets of 16 each in the body cell of a female bee, one set being derived from the male parent and the other from the female parent.

Let us consider for a moment one of the chromosomes in the paternal set. It has its own shape and size and carries its own special set of genes arranged in a particular order. In these respects it is quite distinct from the other chromosomes in the paternal set. There is however, one chromosome in the maternal set which is like it in shape, size, and the number, order, and functions of its genes.

These two like chromosomes make a pair, and being similar in form and size are described as homologous.

The genes of one member of a homologous pair correspond both in number, relative position, and function with those of the other. A gene occupying a particular position in the series of one member of a pair corresponds in function with the gene which occupies the same relative position in the series of the partner.

Thus in any particular pair of homologous chromosomes there are two genes (one in each member) responsible for a particular characteristic in the individual organism. For example one of the genes in a particular chromosome may be regarded as responsible for the colouring of the first three abdominal tergites of the honey-bee. If the bee is of pure race (e.g., "blacks") both of the colour genes in one homologous pair of chromosomes will be responsible for the black colouring of the tergites, and the progeny is said to be homozygous for "black". If however the bee is a hybrid (black crossed by Italian with golden colouring) one of the genes will be responsible for "black" and the other for "golden" and the progeny will be hybrids (p. 139) and described as heterozygous. In every such case there are only two of these particular colour genes in a cell of an individual, and therefore not more than two colours to be transmitted to the tergites of progeny.

At rare intervals a gene changes, and although it remains the same gene and occupies the same relative position in the chromosome it causes a new effect. Such a change is known as mutation. The gene may in fact mutate several times in the course of ages, each time producing a new effect. Thus if we assume that all the races of our hive-bee originated in a wild type and that this was black, then successive mutations of the aforementioned colour gene led to the appearance of the golden of the Italian, the deep bronze of the Cyprian, the grey of the mountain Caucasians, the pale yellow of the Palestinians, and so on.

Such alternative forms of a gene occupying the same locus on a particular chromosome, and originating from a single gene, are known as allelomorphs. The different forms are transmissible from generation to generation in accordance with Mendel's laws (pp. 138 et seq.).

Cell division.

Before proceeding further we may consider very briefly the typical modes of cell division.

(1) *Division of vegetative cells.*

As previously stated growth and repair of tissues is maintained by the repeated division of body cells. During growth the successive divisions take place after short resting stages. The cells of adult organisms divide after much longer intervals, the resting stages sometimes lasting for years.

During the resting stages of a cell the chromosomes usually become indistinguishable from the nuclear sap surrounding them but they become visible again when division is about to take place. At this time each comprises two constituent thickened threads, called chromatids, which lie close together throughout their length.

These separate, and one set of chromatids gradually moves towards one pole of the nucleus and the other set towards the opposite pole. Each set carries the same genes as the other set and becomes a set of daughter chromosomes. The original nucleus, and also the cell surrounding it, divides into two, thus forming two daughter cells each of which encloses one of the daughter sets of chromosomes. Each daughter cell thus contains the same number of chromosomes and the same genes as the parent cell, and therefore the physical characteristics of the individual, as controlled by the genes, remain unchanged throughout its life. This mode of cell division, in which there is a longitudinal splitting of the chromosomes, is known as "MITOSIS".

(2) *Reduction division.*

In Chapter III we have referred briefly to the union of male and female pro-nuclei to form what is known as a segmentation nucleus,—that is to the fertilisation of the egg.

Since each of the body cells of the female bee contains 32 chromosomes and each of the body cells of the drone 16 chromosomes it follows that, if there were no preliminary modification, the union of male and female gametes would result in a segmentation nucleus containing 32 + 16 chromosomes, and that the number of chromosomes in a cell would increase in each new generation.

This is prevented by what is known as the Reduction Division, one of a series of changes involved in the "ripening" or "maturation" of the egg cell. The details of these changes (96, p. *47* et seq.) cannot be given here, but briefly they are as follow:—

(1) The egg is being laid by the queen. Its nucleus contains 16 paternal and 16 maternal chromosomes, appearing as long strands.

(2) Each chromosome divides lengthwise into two strands or "chromatids" which remain held together.

(3) Like chromosomes (one paternal to one maternal) come together in pairs. They lie closely alongside each other and then become spirally intertwined, the two chromatids of the one being wound round the two chromatids of the other.

(4) There is a crossing over or interchange of groups of genes in some regions of the paired chromosomes, so that each chromosome then comprises regions of paternal and maternal genes.

(5) The chromosomes separate, half moving in one direction and half in the opposite direction. They assort themselves at random however so that each new half set contains some paternal and some maternal chromosomes. About each half set (16) of chromosomes a daughter nucleus is formed. This is the *reduction division.*

(6) Each daughter nucleus divides again into two nuclei, the constituent chromatids of its 16 chromosomes separating as in mitosis (p. 127).

(7) Of the four nuclei thus formed within the egg three are absorbed and disappear, leaving one—the "pro-nucleus"—with the reduced number of chromosomes which bear both paternal and maternal characters.

These changes are collectively known as *MEIOSIS.*

The consequence of the processes of crossing over and the random assortment of the chromosomes in the reduction division is that hereditary factors derived from both male and female parents are borne in each ripe germ-cell.

The maturation of the sperm-cell, which occurs whilst the drone is in the larval and pre-pupal stages, (24, p. *106* et. seq.) is similar in some respects to that of the egg, but no reduction of chromosomes is effected. The first polar body which separates from the nucleus contains no chromosomes. The parent nucleus is therefore left with the unreduced number (16) of chromosomes, this number being unaffected by subsequent mitotic division.

The union of the ripened sperm and egg restores the normal number of chromosomes (32) characteristic of the female bee.

The fertilised eggs of a queen therefore, which yield female progeny only, bear unit characters derived from her male and female parents together with those derived from the sperm of the drone with which she mates. As subsequent cell multiplication is effected by mitotic division these unit characters exist in all the body cells of her female progeny.

Polyploids.

In some organisms are body-cells with more than the normal two sets of chromosomes. These arise from failures in cell-division in meiosis, the chromosomes dividing but the nuclei and cells failing to do so. When this happens in a diploid organism each cell affected will contain four haploid sets, or double the normal number of chromosomes, and is described as tetraploid. In this way patches of tetraploid tissue surrounded by diploid tissue may be formed (96, p. *44*). A reference to tetraploid tissue in relation to diploid parthenogenesis is made on p. 157. Cells with multiples of the haploid set of chromosomes are collectively known as polyploids.

Linked characters.

As previously stated (p. 128) there is, at a certain stage in Meiosis, an interchange or crossing over of blocks or groups of genes between paired chromosomes. The grouped characters which cross to one chromosome persist as a group in the body cells of individuals of the first generation which carry that chromosome.

In succeeding generations however the regions of crossing over are not necessarily the same.

On the other hand the genes carried in a particular chromosome constitute what is known as a linkage group. They go into a mating together, in so far as they are not modified by crossing over, and keep together, as shown by the appearance of their characteristics in succeeding generations.

In the breeding of a queen therefore it does not follow that because we have been able to breed true to a particular colouring we have necessarily bred true to other physical characteristics of a parent, for the genes controlling these may be carried on different chromosomes which, assorting independently, may or may not appear together in the same individual. We have however bred true to all the characteristics in the linkage group containing the colour gene in question, except in so far as this has been modified by interchange in crossing over.

Evolution.

All living organisms display a constant tendency to changes in their physical constitution. Some changes are favourable, others unfavourable to survival in the struggle for existence. By natural selection the former are promoted whilst the latter tend to disappear.

In this respect the higher plants and animals which increase by sexual union have advantages over unicellular forms of life which multiply by simple repeated cell division. The former by pooling the characters of several individuals produce a greater variety of combinations of unit characters, the favourable ones of which increase the chances of survival.

Mutation.

Heritable variations leading eventually to new species are caused by changes in individual genes or in their arrangement in small regions of the chromosomes. These changes, known as mutations, are rare, and occur suddenly. A mutation introduces a new unit character which if favoured by environment is perpetuated in the stock but which tends to disappear in the course of natural selection if it is not favoured by environmental conditions. Other causes of variation (22, p. *481*) need not be explained here.

The vast majority of mutations, whilst slight in their effects, are disadvantageous to the individuals in which they occur. They impair the balance of the whole gene-complex which in the course of ages of natural selection has become delicately adjusted to environment, and so render individuals in which they occur a little less efficient in the competition for survival in a particular environment.

A mutation which increases the efficiency of an individual in the struggle for existence, and which therefore tends to be perpetuated and to produce a new race, is extremely rare, and usually ages would elapse before a succession of useful mutations would occur and a new race develop which would succeed in displacing an old one.

Attempts have been made to estimate the frequency of gene mutation. Ford (25, p. *21*) states that

> "probably a given gene seldom mutates in more than one in 300,000 individuals,—and usually more rarely";

and referring to advantageous mutations, (25, p. *46*),

> "on general grounds they might be expected to occur perhaps once in 10^9 individuals.[1] At any rate their frequency may be something of this order".

Seeing that in bees the vast majority of individuals do not have progeny, the frequency of mutations in the queens and drones which mate is enormously decreased as compared with that in other insects, whilst the chance

[1] $10^9 = 1,000,000,000$

of an advantageous mutation occurring in them is almost infinitely small.

In connection with queen-rearing under ordinary conditions therefore the significance of mutations is negligible. Their frequency however is greatly accelerated by the action of radium or X-rays,—according to Ford, citing Oliver, (25, p. *22*) by "at least fifteen to twenty thousand per cent, the rate of increase being proportional to the energy of the dosage absorbed, and independent of the wave length". If therefore a queen be caused to carry on her thorax a small amount of a radio-active substance, as in the case of the device described on p. 320, the question of induced mutations may assume some significance. Practically all the mutations induced would tend to reduce efficiency or viability, but since less than one per cent of the bees and queens bred in a stock—and probably very much less on account of the small dosage of radio-active substance used—would be likely to be affected, the general effect on a stock from the point of view of practical beekeeping would be negligible.

Continued use of the device on successive generations of a strain selected for breeding might however prove to be deleterious and it would be prudent therefore to refrain from applying it to a queen from which young queens or breeding drones were to be reared.

Sports.

A changed gene will be transmitted to succeeding generations, but if its character is recessive (p. 139) its characteristic will not appear in an individual until it accidentally meets a similar gene (of the same or similar descent) in another individual. The chances that this will happen are increased by in-breeding (53, p. *64*). The resulting new individual is popularly described as a "sport". By self-fertilisation in the case of a plant or in-breeding in the case of an animal a sport can be bred true and perpetuated.

In wild life sports have little chance of survival in competition with established types, but in cultivated plants and domesticated animals (e.g., birds) they are helped to survive by artificial selection. A sport which might arise in bees would have a very poor chance of survival for the vast majority of bees, both male and female, have no progeny, and even if by a rare chance, a sport should arise amongst queens or drones which mate it would probably be undetected, and unaided by artificial selection would not be likely to be perpetuated.

Mutations in bees as in other animals may happen at any time. They may remain recessive for many generations and then suddenly appear, or re-appear, as in the case of the lack of pigment in the eyes of albino drones.

Influence of environment.

A gene is said to interact with its environment, which may be external, involving such conditions as temperature and food, or internal, involving the balance of the genes as determining the reaction of the whole gene-complex to a change in any one of its parts (22, p. *480*). Thus a given gene produces a particular effect in its normal environment, but if this is considerably changed the gene will, without changing in itself, produce a different effect. If restored to its original environment the gene will produce its original effect.

During the past 100 years there has been much movement of bees from one part of the world to another and bees so moved to new environments might be expected to show differences in appearance, physique, and behaviour. Michailov (73, p. *626*) who worked on the tongue lengths of bees in the plains of European Russia found that there is a regular increase in the length of the tongue from the north towards the south. He also found that the tongue length of a colony of bees increased as the season advanced; that bees incubated at 35°C. had longer tongues than those incubated at a lower temperature; and that tongue length was greater in bees reared in white than in dark combs. Alpatov (73, p. *626*), whose work confirmed that of Michailov, found that bees reared from underfed larvae had shorter tongues than those reared in normal conditions. These results go to show that bees are susceptible to changes in external environment and that the effects of such changes may not be inconsiderable in bee economy.

Mendelism.

Each fertilised egg contains a double set of chromosomes, one derived from the male and the other from the female parent. Laws relating to the distribution of these as exhibited in the characteristics of resulting individuals were formulated by Mendel in the middle of the last century although he knew nothing about the chromosomes and their genes. These laws however are limited to their application to characters or linked groups of characters which assort in meiosis independently of one another.

Mendel.

Gregor Mendel, (1822–1854), priest, science teacher, and later Abbot in the Monastery of Brünn in Silesia, after several years of patient experiments in the crossing of plants, discovered two fundamental principles of inheritance relating to the transmission of unit characters to offspring, viz., the segregation of contrasting characters, and their independent assortment after fertilisation. Later these principles were shown to apply to animals as well as plants.

Mendel's conclusions, published in 1865, were almost ignored and were then forgotten until rediscovered in 1900.

We are told (6, p. *327*) that he experimented with bees, having in his apiary about 50 stocks of different races. His notes on these have not been found and it is believed that they were destroyed. He attempted to control mating by netting his own room with gauze, but failed. The impossibility of securing desired mating in the circumstances must have vitiated his experiments and it would not be surprising if he himself destroyed his records.

Most of Mendel's experiments were made with garden peas in the monastery garden at Brünn. He conceived that plants must be complex organisms containing unit characters which could be transmitted in certain combinations to their offspring.

By systematic controlled pollination he bred pure and hybrid plants from parents having distinctive physical characters such as tallness and shortness, smooth and wrinkled seeds, and green and yellow cotyledons. These pairs of characters are allelomorphs (p. 125) and from patient observation Mendel was able to show that they are transmitted to descendants according to definite rules (Pl. XIII, A).

The characteristics of bees which are of special interest to the queen-breeder are:—

1. Colour e.g., Golden. Black.
2. Prolific. Unprolific.
3. Productive. Unproductive.
4. Docile. Irritable.
5. Swarming. Non-swarming.
6. Long-lived. Short-lived.
7. Long-tongued. Short-tongued.
8. Prone to robbing. Little inclined to rob.
9. Disease resisting. Susceptible to disease.
10. Using much propolis. Using little propolis.
11. Making dry (white) comb cappings. Making wet (dark) cappings.

Of these characteristics we may make the following remarks:—

(1) Contrasting colours of the first five abdominal tergites serve as evidence of parentage. Since

their transmission complies with Mendel's laws they are considered to be controlled by unit genes.

(2) The Mediterranean races are more prolific than the dark races.

(3) Every experienced beekeeper is familiar with the great variability shown by different strains within the same race and by individual stocks within the same strain, with respect to the amount of honey they store in excess of their requirements, even when they are approximately equal in strength, situate in the same environment, and subject to the same management.

(4) The differences in temper of bees are well known. Carniolans and Caucasians are gentle, as are some strains of Italians and English blacks, whilst Cyprians and Syrians and some hybrids may be implacable.

(5) Some strains of bees show little inclination to swarm whilst others swarm to excess. Carniolans, Dutch blacks, and the German heath bees are inveterate swarmers.

(6) The individual bees of the prolific races are considered to have shorter lives than those of the races whose economy demands more restricted breeding. Some writers hold the view that there is some correlation between the length of life of a queen and that of her worker progeny. Whether this be so or not stocks of bees of the same race appear to vary in respect to the longevity of individual bees. Miss Betts (10, p. *102*) reviewing an article in Deutsche Imker, August 1934, describes experiments of E. Storch in this connection. He marked newly-emerged bees of two colonies—"the one a good honey producer, the other given to overmuch breeding and little surplus". In the inferior stock the first diminution of marked bees began on the thirtieth day

and the last of them disappeared on the thirty-ninth day. In the good stock disappearances began a week later and the last marked bee disappeared on the forty-eighth day. Storch attributed the longer lives of the latter bees to increased flying power and efficiency.

It may be observed that the stock with the longer-lived bees would tend to remain strong, gather more nectar—thus limiting the size of the brood nest—reduce the need for renewal of the population, and so prolong the useful life of the queen.

(7) From extensive studies of the tongue lengths of bees Chochlov (73, p. *624*) concluded that each of six races of bees examined by him has a specific tongue length. His work was confirmed by Alpatov who found that Italians have longer tongues than blacks, and Caucasians the longest tongues of all.

(8) Italian bees are more addicted to robbing than the black bees.

(9) Efforts to breed disease resisting stock, especially in relation to American Foul Brood, have met with but limited success (73, p. *293*).

(10) Caucasians use larger quantities of propolis than bees of other races.

(11) The dark races of bees leave small air spaces under the cappings of their honey cells which gives these a white appearance. Italians do not as a rule leave this space. Their cappings are in contact with the honey and consequently have an irregularly stained appearance.

Since the characters specified above include those which are desirable in bees the question as to how far they are transmissible in successive generations is of special interest to the queen-breeder. They may be grouped as follows:

Group 1. Unit gene characters.
Group 2. Multiple gene characters.

Group 1. The colours of the dorsal abdominal tergites, particularly of the first five; of the bands of hair upon them; and of the scutellum, are probably controlled by unit genes. They serve to indicate parentage and their transmission is observed to be in accordance with Mendel's laws.

Group 2. All the remaining characters, with the possible exception of No. 7 fall into this group. Each is dependent on a complex of elements of the physical constitution controlled by many genes. Those which involve behaviour depend on the structure of the nerve system which also is probably controlled by many genes.

Where many genes are involved some will be on separate chromosomes, and since these assort independently in meiosis (p. 128) all the characters controlled by the genes will not necessarily go into one offspring. Crossing over (p. 128) may further complicate the transmission.

These multiple gene characters therefore, although found by experience to be in some measure hereditary, are subject to variations, and the breeder who desires to perpetuate them can do so only by culling the queens of stocks which do not show the desired qualities.

In the ensuing explanation of Mendel's principles we refer for convenience to allelomorphs for colour, viz., yellow and black, as being easily observable. The theory is equally applicable to other allelomorphic characters.

A fertilised egg contains corresponding genes derived from both parents. If these parents are pure-bred in respect of a particular character, e.g., if they have the characteristic colour of their race, the colour-producing genes in the fertilised eggs will be alike and all the progeny will resemble the parents in respect of colour.

If the parents are different in respect of the genes responsible for their characteristic colours, both of the unlike genes are present in the fertilised egg, and the immediate progeny (hybrids) will be either,—

(1) like one of the parents,
or (2) intermediate between the two.

In the former case the colour of one parent appears in all the offspring to the exclusion of the other. The one which appears was called by Mendel the "dominant" character of the pair and the one which does not appear the "recessive" character. The latter however is only latent, for both characters re-appear in succeeding generations according to definite laws.

Breeding from a pure race of bees.

Let us now consider what happens when a pure-bred yellow or "golden" Italian queen mates with a drone of the same race.

(We may here premise that we cannot always be certain that we are dealing with pure-bred races. Apparently-pure Italians or Blacks may carry recessive characters which may suddenly appear in the progeny and which may give rise to misleading deviations from the results expected. By "pure-bred" therefore we postulate that no deviation from the normal has been observed for a considerable number of generations.)

Each egg-cell after the reduction division (p. 128), and each sperm-cell contains a chromosome bearing the gene responsible for the gold colouring of certain of the abdominal tergites so that every fertilised egg contains two G-bearing[1] chromosomes—one derived from each parent—which may be denoted by GG. Consequently the colouring of the resulting progeny

[1] In this connection G and B are used as the initial letters of "golden" and "black". Y for "yellow" is avoided because it is used in another connection (p. 123).

will be golden like that of the parents. Pl. XIII, B. F4 (a).

Similarly when a pure-bred black queen mates with a black drone of the same race the colour-bearing chromosomes of the progeny will be BB, and the colour of the progeny black, Pl. XIII, B. F4 (b).

In each of these cases the colouring of the drones will correspond to that of the female bees.

Crossing of differently coloured pure-bred races.

If a pure-bred golden queen is mated to a black drone every egg-cell after the reduction division contains a G-bearing and every sperm-cell a B-bearing chromosome. The fertilised eggs will be G- and B-bearing and will produce "hybrids". These correspond to Mendel's Family 1. (Pl. XIII, A, F1).

The colouring of these hybrids depends on whether yellow or black is dominant. Generally it has been found that when yellow and black races have been inter-bred, yellow is dominant (73, p. *90*). Sladen (76, p. *60*) extracted from crosses between English Blacks, Italians, and American Goldens a hybrid with yellow bands on the 2nd to the 5th abdominal rings which he named "British Golden". Roberts (70, p. *259*) in his experiments found that five-banded yellow queens artificially inseminated with sperm from black drones produced only progeny of the intermediate class, i.e., with three instead of five yellow bands.

The female individuals of the first family from inter-bred Italian and black parents may therefore be expected to be intermediate in colouring and to resemble the yellow rather than the black parent. The drones will inherit colour only from the mother (Pl. XIII, B, F1).

Inter-breeding within hybrid stock.

Mendel experimented with plants with male and female diploidy. His ratios of character inheritance do not therefore apply directly to bees, the males of which are haploid (p. 123).

Maisonneuve (62, p. *39*) in his short account of Mendelism in relation to bees takes no account of the haploidy of the drone and consequently his explanation and diagram are not directly applicable to his subject. Sladen (76, p. *56–57*) is more cautious. He illustrates the Mendelian scheme of inheritance from crossed diploids but omits to say that this does not apply to bees.

In Plate XIII are two diagrams, A and B, the former showing the Mendelian scheme of inheritance from pure but unlike diploid parents, which applies to most plants and animals, and the latter showing inheritance from pure but unlike parents, one diploid and the other haploid, which applies to bees.

In A two bars may be taken to represent the two colour genes of the diploid individual whether male or female. In B two bars represent the colour genes of the queen-bee and a single bar the one colour gene of the drone.

In each case also, for the sake of simplicity, it is assumed that progeny from a single mating comprises only four individuals. In A, relating to most plants and animals, these comprise both males and females. In B, relating to bees, they comprise two queens and two drones.

We may now consider, with references to the diagram B in Pl. XIII, what happens if we attempt to perpetuate a hybrid strain by interbreeding, eliminating pure queens as they occur.

We first premise that after maturation each egg-cell nucleus and each sperm-cell nucleus contains either a G-bearing or a B-bearing chromosome. If a B sperm fertilises a B egg the resulting bee will be BB, (black). If a G sperm fertilises a G egg the resulting individual will be GG, (golden). If a B sperm fertilises a G egg, or a G sperm a B egg, the resulting individual will be

PLATE XIII

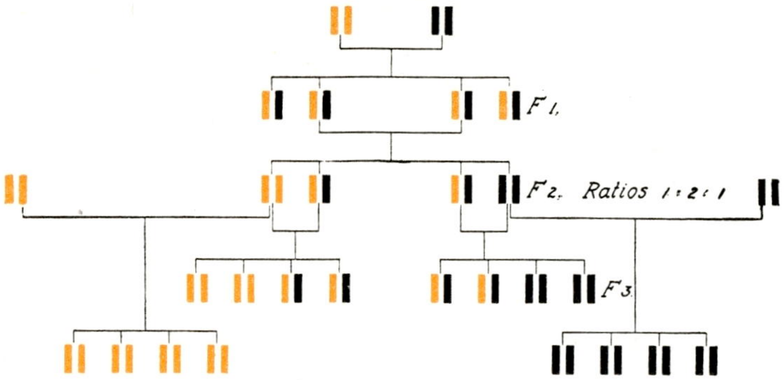

A. Diagram showing Mendelian scheme of inheritance from cross-mated diploid parents.

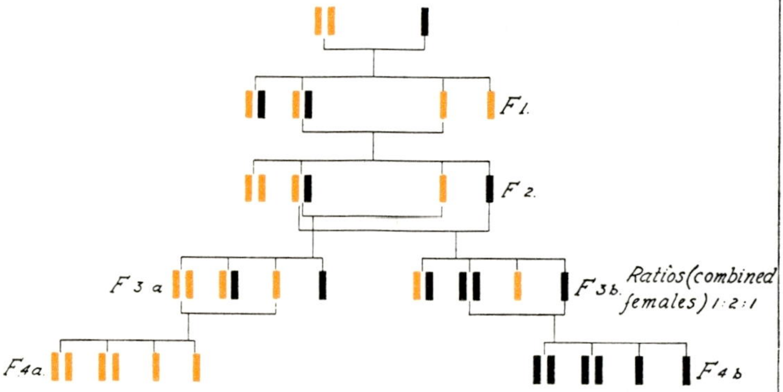

B. Diagram showing scheme of inheritance from cross-mated BEES:— females diploid, males haploid.

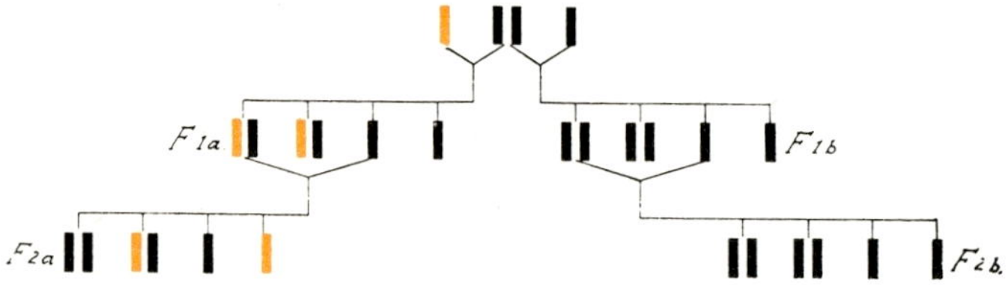

C. Diagram showing effect of plural mating on drones of the first and second generations

SCHEMES OF INHERITANCE

BG or GB (which are the same), i.e., hybrids. Owing to parthenogenesis (p. 124) the drone progeny of a hybrid queen will be of two kinds in approximately equal numbers, some derived from B-bearing, and others from G-bearing unfertilised eggs.

From Diagram B, F1 it is apparent that the union of a pure "golden" queen with a black drone produces hybrid females and "golden" drones, the latter being derived only from the queen mother.

The union of one of the hybrid queens with one of her half-brother "golden" drones in F1 yields as progeny the family F2 in which half of the queens are pure "golden" and half hybrids, and drones, half of which are goldens and half blacks, each drone inheriting one of the colour genes of the mother.

Eliminating the pure queens in F2 we see that there are two possible kinds of union, viz.:—

(*a*) A hybrid queen with a "golden" drone.
(*b*) A hybrid queen with a black drone.

In the progeny from union (*a*) shown at F3 (*a*) queens will be in the ratio of 1 golden: 1 hybrid, or GG: GB.

In the progeny from union (*b*), shown at F3 (*b*) queens will be in the ratio of 1 hybrid: 1 black, or GB: BB.

In each family black and golden drones will be produced in approximately equal numbers, and we may therefore assume that in a very large number of matings of hybrid queens golden and black drones will be involved in approximately equal numbers.

If therefore we add the two typical families in F3 together as representing the combined progeny from large numbers of matings of hybrid queens in the apiary, we get as types of queen progeny, in approximately equal numbers, GG+GB+GB+BB= GG+2GB+BB, i.e., the three types in the ratios 1: 2: 1.

These ratios are the same as for the Mendelian family F2, derived from diploid hybrid parents, (Diagram A, F2).

Both diagrams show how pure types may be derived from hybrids. In B, two pure families of the fourth generation are shown. Other possible families of this generation would be similar to F3 (*a*) and F3 (*b*).

Crossing of hybrids with pure races.

This really amounts to the same thing as described in the preceding paragraph since only the queen, being diploid, can be the hybrid, and the drone, being haploid, must be of a pure type. The progeny are as shown in B, F3 (*a*) for a golden drone, and in B, F3 (*b*) for a black drone. In each case half the female progeny will be hybrids and half of a pure type, whilst the drone progeny will be pure and of two types.

In the diagrams we have assumed that only four descendants proceed from each union of two parents. In the case of bees of course the immediate descendants from a single union may number hundreds of thousands. The ratios of 1 : 2 : 1 for the females in the Families 3 (B) apply the more accurately as the number of individuals, or in the case of queens, the number of mated queens is increased.

It appears from the diagrams that by isolation and interbreeding we can rear pure Italian or pure Blacks from hybrids. This is true in respect of colour and the other characters which may be in the same linkage group. (p. 130). In respect of other characters, which may be included in other linkage groups, the descendants may differ from their pure-bred ancestors.

Uncertainties of breeding within hybrid stock.

The presence of differently coloured bees in a stock derived from a hybrid queen is familiar to most beekeepers. The owner of such a stock, which may have shown excellent qualities, naturally desires to perpetuate his strain, but when he selects eggs or young larvae for queen-rearing it is impossible for him to determine whether they will produce individuals of the GG, GB, or BB types. By a rare chance he may even select larvae producing all goldens or all blacks to the exclusion of his hybrids. Only by severe and continued culling can he preserve his hybrid strain and this would certainly prove unprofitable to a queen-breeder.

His difficulties would be increased moreover if, instead of breeding for colour only, he selected his queens having regard to two allelomorphic pairs of characters,—e.g., golden as contrasted with black, and long-tongued as contrasted with short-tongued. (We assume, for the purpose of this consideration, that tongue length is a unit gene character). The types of progeny he should obtain are shown in the diagram on p. 146.

In this diagram we denote yellow by G and black (lack of yellowness) by g and assume that G is dominant to g. Similarly we denote long-tongued by T and short-tongued (lack of long tongue) by t, and assume that T is dominant to t. We also assume that each female gamete and each male gamete carries one of each pair of the genes responsible for the allelomorphic characters, which assort independently. We further assume that we are dealing with average results of large numbers of matings within the same hybrid stock (pp. 141–143).

Each male gamete (sperm) and each female gamete (egg) will bear one of the combinations GT, Gt, gT, gt.

If in the following diagram we set out the characters of the male gametes along the top line, and repeat them vertically, and those of the female on the lower lines of the left hand column, repeating them horizontally, we show all their possible combinations.

Each square may be considered to represent a fertilised egg carrying the two pairs of genes in question.

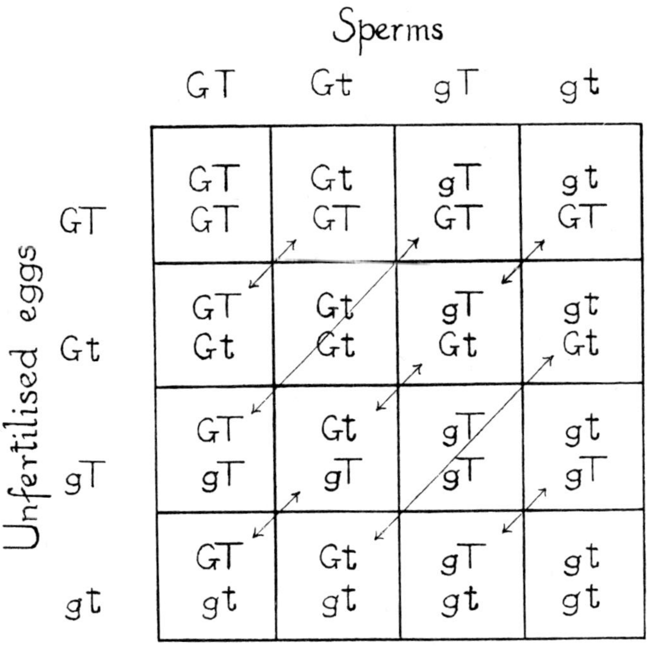

Of the 16 combinations shown those connected with arrows contain the same characters and therefore are identical. Nine are different from one another, viz:—

GT,	Gt,	gT,	Gt,	g t,	g t,	gT,	g t,	gt
GT	GT	GT	Gt	GT	Gt	gT	gT	gt.

Of these four have both dominant characters and one has neither.

Thus from unrestricted breeding within hybrid stock

and having regard to two pairs of allelomorphic characters we obtain 9 types of female bees, or if we are rearing queens from them on a large scale, 9 types of queens.

If breeding within hybrid stock were carried out with regard to three allelomorphic pairs of characters the number of types produced would be considerably increased.

From these considerations it is evident that, even if we can secure isolation, continued breeding within a hybrid strain with a view to perpetuating two or three specific characters is an almost hopeless task. The queen-breeder who desires to produce stock with constant characteristics should therefore cultivate one of the pure races. As Phillips (64, p. *455*) points out,—"The first cross is often desirable from the standpoint of honey gathering, but it is better to breed from pure stock only, for the offspring of a hybrid queen is exceptionally variable, and it is a matter of chance if good stock results."

Tendency of hybrids to revert to pure types.

As previously shown (p. 143) the combined female offspring of a number of hybrid queens bearing colour genes for B and G, half being mated to B-bearing and half to G-bearing drones, will in each generation comprise individuals in the ratios 1BB: 2BG: 1GG, i.e., 1 golden: 2 hybrids: 1 black.

If we assume that we can secure complete isolation; that we breed in the first instance from a hybrid queen and a drone of the same maternal descent; that each pure queen mates with a drone of her own type, and that from each union four typical daughter queens are produced it will be seen from the following table, adapted from Mendel (6, pp. *349–350*) that the pure types increase much faster than the hybrids,—the proportion of the latter constantly diminishing but never disappearing.

Generation	Descendants GG	GB	BB	Ratios. GG	: GB	: BB
1	1	2	1	1	2	1
2	6	4	6	3 :	2 :	3
3	28	8	28	7 :	2 :	7
4	120	16	120	15 :	2 :	15
5	496	32	496	31 :	2 :	31
n				2^n-1 :	2 :	2^n-1

In the 10th generation, for example, 2^n-1 will be 1023, and therefore of every 2048 female descendants 1023 will be golden, 1023 blacks, and 2 hybrids.

It may be said that in practice the pure types would inter-breed with the production of more hybrids, but these in turn would continue to produce an increasing proportion of pure forms.

In this country climatic conditions are less favourable to the economy and survival of the yellow than of the indigenous dark bees mainly on account of the greater prolificacy of the former and their consequent poverty after poor seasons. Of the pure types derived from hybrids therefore the black tend to survive and the yellow gradually to disappear.

In view of these tendencies we can well understand why our bees, the importation of queens from Italy having ceased, are gradually reverting to the original dark types of the country, whilst the proportions of yellows and hybrids are diminishing.

In-breeding.

In-breeding is the system by which the fixation of desirable characters is sought by the mating of closely related individuals. In out-breeding increased vigour is sought by the mating of unrelated individuals.

It is a popular notion that in-breeding is necessarily harmful, but as Crew and Lamy (21, p. *107*) remark, "there is nothing good or bad in in-breeding itself.

It is good or bad according to whether the individuals exposed to it are good or bad."

In the case of bees the possibilities of in-breeding are limited partly on account of the difficulty of selecting the male parent, and partly because it is impossible to mate the female with her father or son. She may be mated with her parthenogenetically produced half-brother or, since several generations of a stock may exist at one time, with her half-uncle, half-nephew, or half-cousin. The term "half" is used here because the queen inherits from two parents but the drone from only one.

In in-breeding any two mated individuals should possess the characters it is desired to perpetuate and these characters will then appear in some of the progeny. But the same individuals will also possess many recessive characters, some of which are likely to be disadvantageous. Whilst these are obscured by the dominant characters their bad effects do not appear.

Since however members of the same family possess the same genes the chance of two individuals meeting, each with the same deleterious recessive gene, is much greater than when unrelated individuals with different sets of genes are mated.

The double dose of a deleterious recessive gene will produce an undesirable effect in the progeny with consequent deterioration of the stock. If however the recessives carried by the parents are not harmful no evil results will come from in-breeding.

The in-breeder therefore preserves progeny showing the desired characters and eliminates the rest. Continued in-breeding without such selection generally leads to serious deterioration (21, p. *101*).

Out-breeding is resorted to when it is desired to introduce a new character or increased vigour to a stock. The increased vigour is apparent and fairly uniform in the first (F1) generation but in the next (F2) generation there is much variation and the "hybrid vigour" is dispersed and tends to be lost (21, p. *102*). Out-breeding in animals is therefore usually practised to produce the first generation, but this is not used for further breeding.

The queen-breeder therefore who desires uniformity in the qualities of his stock should not breed from hybrids, in the descendants of which much variation occurs. He should avoid out-breeding unless with bees of the same race and then only if he follows up with a course of in-breeding. Generally he should practise in-breeding within one of the pure races, carefully selecting the best and at the same time consistently eliminating undesirables.

Apparent deviation from parthenogenetic inheritance.

It has already been shown that the drone has no father. Developed from an unfertilised egg, it cannot have genes—and therefore the characters for which these are responsible—derived from the drone which mated with the queen-mother, and therefore the characters it inherits can be only those transmitted by its mother. It is commonly accepted therefore that the drone progeny of a queen of pure race, however mated, will be of the same variety as that queen, and may be utilised accordingly for breeding.

Doolittle (23, pp. *73–74*) however writes at some length to show that "drones are contaminated, to a certain extent, by the mating of a queen of one blood with a drone of another blood" and that subsequent female progeny show evidence of such contamination. He concludes by saying that "if such ('contaminated') drones are allowed to fly in your yard you cannot expect any satisfactory degree of purity from queens reared therein". Other observers have expressed similar views but have not been able to adduce a scientific reason for this apparent deviation from parthenogenetic inheritance.

In seeking for a reason we should bear in mind

(1) The doubtful purity of American queens which may be carrying recessive colour genes derived from unknown ancestry.
(2) The possibility (extremely unlikely) of a mutation.
(3) The effects of plural mating.

In respect of (3) we must remember that in addition to the popular Italian bees,—Blacks, Caucasians, and Carniolans are extensively used in the U.S.A.; that multiple mating is common; that there appears to be a marked tendency for queens to mate with drones of different races if these are available; and that a queen mated with two drones of different races yields two kinds of progeny (70, p. *255–9* and *303*).

Diagram C, Pl. XIII, illustrates the effect of the double mating of a queen in the first and second generations. Suppose that in an apiary of all black bees a black queen mates with a black drone and that, unknown to the bee-keeper, she also mates with a stray Italian drone. She produces two types of female progeny as shown in F1 (*a*) and F1 (*b*), the former comprising hybrid queens and black drones, and the latter all black queens and drones. The drones of both types are blacks because they inherit colour genes only from their mother.

In the next generation, F2, the union of one of the hybrid queens with any one of the drones yields queens half blacks and half hybrids, and drones half blacks and half Italians F2 (*a*). If we assume that half the eggs of the grandmother were fertilised by sperms of one drone and half by sperms from the other and that the average of the results of a large number of matings in the first generation is taken, the black drones of the second generation will be to the Italian drones in the ratio of 3 : 1.

It is easy to see from the diagram that more hybrid queens and Italian drones would appear in the third and succeeding generations, all descended from the original black queen assumed by the beekeeper to have been mated with a drone of her own race. This would account for the progressive "contamination" observed by Doolittle. The undesired drones first appear in the second generation F2(*a*) and are attributable to the plural mating of their grandmother. The drones of the first generation may therefore be safely used for selective breeding but not those of succeeding generations.

These considerations emphasise the need, if we seek improvement by systematic selection, for breeding only from one of the pure races, and of securing the maximum of isolation.

Laying workers.

Most beekeepers are familiar with the condition of a stock of bees containing laying workers. These appear during the summer months when the stock has remained without a queen and young brood for two or three weeks or longer. It is believed that they ingest or absorb some of the accumulating brood-food which they secrete, and which would ordinarily be needed by the queen and brood. This causes their ovarian tubules to develop and produce eggs.

The eggs are deposited promiscuously in both worker and drone cells, at first singly, and later, when the number of laying workers has increased, in considerable numbers in each cell. The single ones which hatch develop into drones. These are of normal size if reared in drone cells but smaller if raised in worker cells, the prominently raised cappings and patchy distribution of the latter being an indication of the presence of the laying workers (Pl. XIV). If temperature falls low, or the food supply fails, or if young brood needing to be fed is inserted in the hive, laying is discontinued.

Aware of their queenless and broodless state the bees attempt to raise queens over some of the developing drone brood, enclosing the larvae in full-sized but rather smooth-sided queen-cells, and feeding them

[*Photo* L.E.S.

LAYING WORKERS WITH THEIR SCATTERED DRONE BROOD AND A PSEUDO QUEEN-CELL liberally with royal jelly. But these die in the cells before reaching the adult stage, and unless the stock is provided with a queen it is doomed to early extinction.

The production of a drone from a worker's egg is similar to that from the unfertilised egg of a queen. Both worker and queen are diploid and their eggs after the reduction division are haploid. As the worker has no functional spermatheca, her eggs must normally remain haploid and produce drones. Watson and Whitney (93, p. *155*) have recently proved, by instrumental insemination, that the drones produced from laying workers are fertile.

Diploid parthenogenesis.

So far we have considered the haploid parthenogenesis of the drone. There are many instances in nature, particularly in some species of aphides, where unmated females lay eggs which produce females. In such cases a generation produced by sexual union is followed by several generations of parthenogenetic females, and these by another generation of males and females, and so on. In the eggs produced by the unmated females there is no reduction of chromosomes and they remain diploid (53, p. *204*).

For many years it has been known that diploid parthenogenesis occurs in some of the Hymenoptera, including bees. Morgan (53, p. *237*) refers to rare appearances of sexual females from the eggs of worker ants, and also to their production by laying workers of the "Cape" bees. Whyte (97, p. *75*) in a communication to the Bee World in 1922 referred to a presentation to the British Museum in 1892 of a queen raised from a laying worker's egg.

Recently (1943) Dr. Otto Mackensen (48, pp. *465–467*) of the Southern States Bee Culture laboratory, University of Louisiana, has investigated the presence of female bees in the progeny of unmated queens. After emergence queens were kept from mating by confinement in cabinets under suitable conditions until they were about 25 days old. Their wings were then clipped and they were introduced to nuclei containing only worker bees, Italian virgins being introduced to Caucasian nuclei and Caucasian virgins to Italian nuclei, so that their progeny could be identified by colour. Excluders prevented the egress of the queens and also the intrusion of drones. When these queens began to lay therefore it was certain that they had not been mated. Most of them started laying when from 30 to 40 days old. Of 50 such queens under observation 21 produced brood containing some worker pupae (not more than one per cent). By grafting the young larvae into artificial queen-cells Mackensen succeeded in raising six queens from 710 larvae grafted from the brood of one queen, i.e., 0·85 per cent. In another case six parthenogenetic queens were reared from a golden Italian. One of these daughter queens was prevented from mating and from her brood a parthenogenetic daughter queen was raised. As Mackensen observes, "The fact that two parthenogenetic queens were obtained indicate that such queens are probably normal diploid individuals."

The theory that such individuals are produced from stolen eggs is now generally discounted, and in reference to it Mackensen remarks that when a stock of South African bees becomes queenless laying workers and parthenogenetic females are produced in great numbers, and that it can hardly be maintained that all of the latter arise from stolen eggs.

As all the eggs produced by an unmated queen are unfertilised they are in this respect similar to those laid by laying workers, and in both cases parthenogenetic females are known to be produced.

The production of parthenogenetic females in bees may be due to the suppression of the reduction division in the maturation of an occasional egg (p. 128) or more probably, as Mackensen considers, to the presence of

PLATE XV

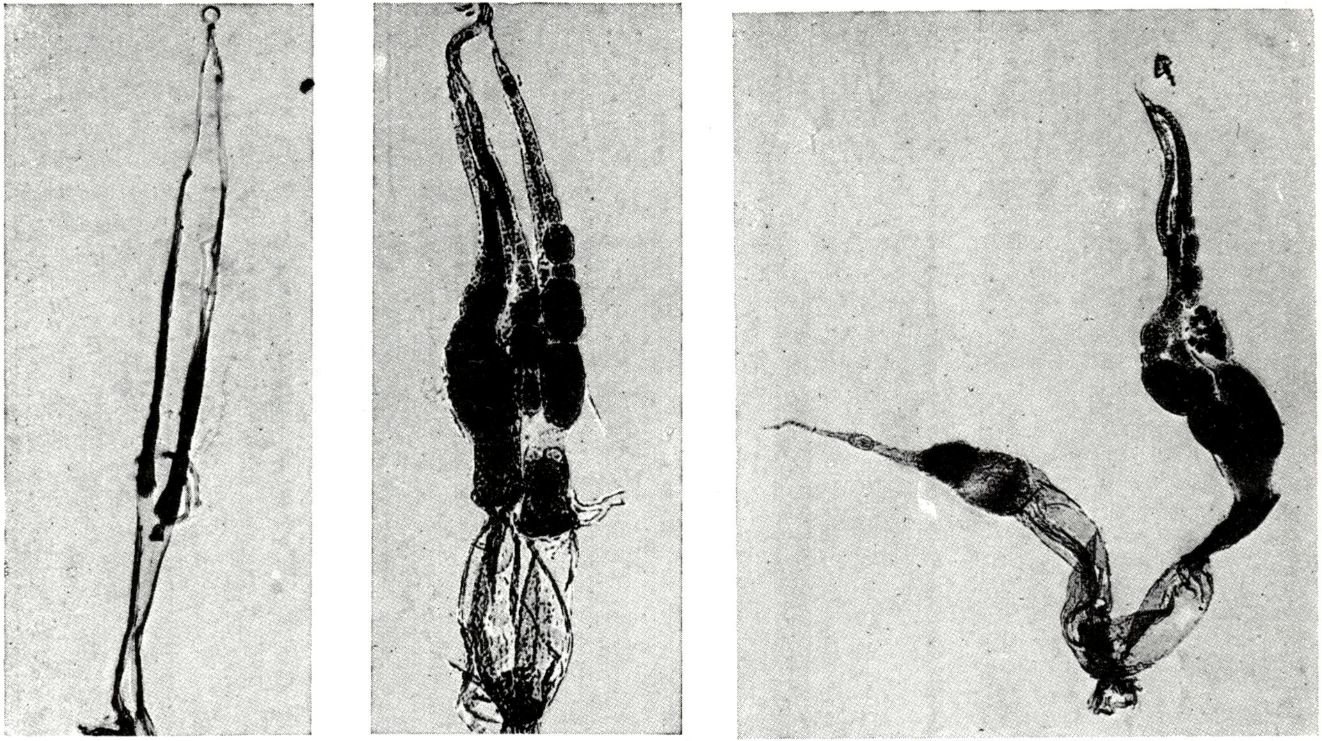

B C [Photos: R. Bassindale

OVARIES OF WORKER BEES

A. Of ordinary worker. *B. and C.* Of laying workers, showing eggs.

tetraploid tissue (p. 129) in the ovaries of the mother.

This tissue would yield tetraploid germ-cells which in meiosis would become diploid and therefore female producing without fertilisation.

Gregg, in a carefully reasoned article in the American Bee Journal of October 1945, attributes the production of parthenogenetic females to the mating of over-fed and therefore over-developed workers and cites two cases where apparent union between drone and worker has been observed. Such union could be only imperfect and a suggestion made is that an egg might be fertilised in the oviduct of a worker-bee—presumably without the functioning of a spermatheca—as a result of such imperfect union.

This explanation, although possibly applicable to fertile workers' eggs, obviously does not apply to those of a definitely unmated queen as in Mackensen's experiments, nor to a drone-breeding queen which had been unrestricted during her period of virginity. In the case of the latter, union with a drone, however imperfect, might be expected to result in the production of far more than 0.85 per cent females.

Bassindale (4, p. *12*), referring to these alternative explanations of diploid parthenogenesis remarks, "It is quite possible that both mechanisms operate either in different colonies or even in the same colony."

The production of female bees from the eggs of fertile workers may well account for,—

(1) The sudden and unaccountable appearance of a queen-cell in a portion of a hive from which the queen is excluded.
(2) The supposed stealing of an egg by a queenless stock for the purpose of raising a queen.
(3) The apparently futile efforts of a laying worker stock to rear queens by raising queen-cells over their drone brood.

In (1) the egg is probably laid by one of the laying workers which are at times present in queen-right stocks.

In (2) the egg is almost certainly that of a laying worker.

In (3) the bees would be quite unable to know which eggs would produce females and the chance that they would select one of these for a queen-cell and so requeen their stock would be extremely remote. Even so we may see in this a provision of nature by virtue of which the efforts of a laying-worker stock to requeen itself are not entirely hopeless.

Sex determination in bees.

On p. 124 we have referred to female diploidy and male haploidy in explaining parthenogenesis in bees, but have assumed the existence of sex ("X") chromosomes. White (1945) however observes (95, p. *271*) that sex chromosomes cytologically distinguishable as such probably do not exist in the Hymenoptera and (p. *267*) that sex differentiation in these, as in certain other orders of insects, appears to be determined by haplo-diploidy, the sex of an individual depending simply on whether the egg is fertilised or not. A corollary to this hypothesis is that there is no sex linkage in bees.

In the wasp Habrobracon a special XY mechanism co-exists with the normal haplo-diploid system of sex differentiation. This abnormal mechanism has been investigated by Whiting (1935), Speicher (1934), Torvik-Greb (1935), and Bostian (1936), (22, pp. *378–9*). It produces XY females, and occasionally, by certain crossings and inbreeding, diploid (biparental) males, diploid (impaternate) females from unfertilised eggs, and triploid females from matings of normal females and diploid males (81, p. *401*). The diploid males lead to reduced fertility (22, p. *378*). It is not known that this system functions in any others of the Hymenoptera (95, p. *275*) and the assumption recently made (91, pp. *15–17*) that it is the mechanism responsible for sex differentiation in hive-bees appears to be unwarranted.

PART II

CHAPTER VIII

UTILISATION OF NATURAL QUEEN-CELLS

In nature the renewal of queens is mainly effected through natural swarming, and the beekeeper who allows bees to swarm freely obtains a proportion of young queens each year. If he does not destroy the older queens these are sooner or later displaced by supersedure.

Reference has already been made to the ideal conditions in which natural queens are reared under the swarming impulse and to the view held by some that "these are the best". Considerations of parentage and mating however conflict with this generalisation.

Our purpose now is to consider how we may best utilise the queen cells raised under the swarming impulse.

With reference to natural swarming we must bear in mind that the urge to swarm resides mainly in the bees that have learned to fly, and that after a swarm has issued, the swarming impulse persists in the parent stock—comprising bees of all ages—until casts have issued with young queens, or, if unfavourable weather supervenes, until all the young queens, save one, have perished in combat. To prevent bees of the parent stock from again swarming with the young queens it is essential to deprive it of its flying (older) bees just before the queens emerge from their cells. This is usually done by the Heddon method of preventing "after" swarms. The swarm is hived close to the parent stock and seven days later,—or earlier if the swarm has been delayed by bad weather,—that is to say, just before the expected emergence of the young queens, the parent stock is removed to a new location. Its flying bees re-inforce the swarm and the weakened parent stock consisting only of young bees, loses the swarming impulse.

It is not uncommon to read in bee literature the advice—lightly given—to divide the parent stock into several nuclei, each provided with one queen-cell, without the essential direction concerning the timely abstraction of the flying bees. If these are allowed to remain in the nuclei the young queens are likely to lead off swarms soon after they emerge, or later if delayed by unfavourable weather. In the case of imposed orphanage (p. 176), the abstraction of the flying bees is not necessary.

That the flying bees of a stock which has prepared to swarm carry the swarming impulse can be shown by the following experiment:—

Two prosperous stocks, A and B, with abundant room in supers, stand side by side. A has raised queen-cells in preparation for swarming. B shows no sign of swarming. Remove A to a new location. Its flying bees join up with B. Weather being favourable, B will then prepare to swarm.

The problem of utilising natural swarm queen-cells is twofold, depending on whether in the stock concerned

(A) is about to swarm.
(B) has already swarmed.

In A we have to prevent both the parent stock and each nucleus from swarming. In B there is a minor risk of swarming, confined to the nuclei.

We now consider the procedure in each case.

A. *When a stock is about to swarm.*

Let us assume that a strong stock, one with desirable characteristics, has been found to be raising queen-cells, that we wish to rear a number of queens from it, and at the same time to avoid the risk of swarming.

(If we do not desire to prevent the issue of a swarm there is no point in disturbing the stock at this stage.)

The following courses are open to us:—

Method 1.

(*a*) Just before the most advanced queen-cells are capped over—not later—remove the queen, two combs of sealed brood without queen-cells, and a comb containing honey and pollen, all with adhering bees, and put them into a nucleus hive. Place this nucleus about six feet from the parent stock, facing in the same direction. As there is an occasional tendency for such a nucleus to swarm on the day after being deprived of its queen-cells, cover the hive entrance with a small piece of excluder zinc. This nucleus will lose its flying bees and will probably desist from swarming.

In the parent stock substitute empty combs, or foundation, for the combs removed. This stock will now contain all the flying bees and queen-cells, and should retain the supers, if any. The queen-cells will be completed and kept warm by the full force of the bees, which would not be the case if the stock were divided into nuclei at this stage.

(*b*) Two days before the young queens are due to emerge from their cells, i.e., five days after the queen-cells are sealed, (the date should be known), take from the parent stock two or three nuclei (p. 283). See that each nucleus has only one queen-cell together with brood and stores. No queen-cells should be left in the parent stock. Place these nuclei near the parent hive as shown in Pl. XVI. Two days later examine each nucleus to make sure that the bees have not begun to raise additional queen-cells. If they have, destroy them, leaving only the

PLATE XVI

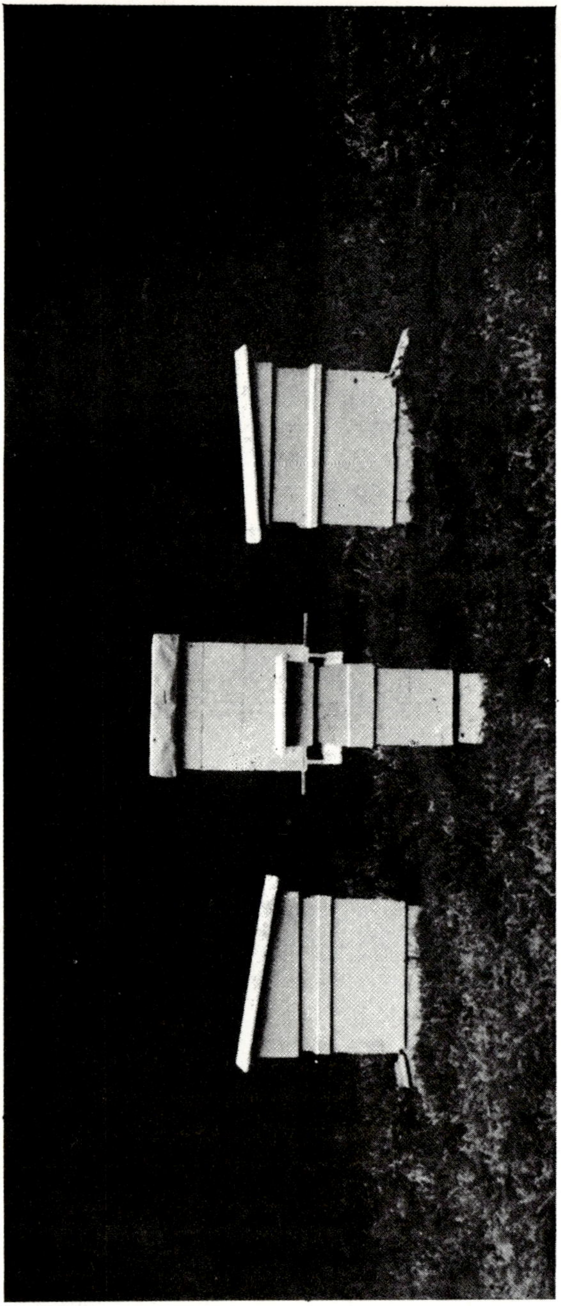

[*Photo: L.E.S.*

UTILISATION OF NATURAL SWARM QUEEN-CELLS. POSITIONS OF NUCLEI original one from which the young queen is likely to be emerging. The nuclei will lose most of their flying bees which will return to the parent hive.

Five or six days after the formation of the nuclei they should be moved to new locations if they are reasonably strong. This will cause them again to lose flying bees to the parent stock just before the young queens may be expected to take their first mating flights. The risk of "mating" swarms (p. 282) will thus be eliminated.

(*c*) The original queen and her nucleus should be restored to the parent hive immediately after the nuclei have been formed. Any queen-cells it may have developed should be destroyed. The stock will now contain all the flying bees and should be strong and capable of yielding a good surplus of honey. It will however probably continue to raise queen-cells, but may be prevented from swarming by one of the methods described in the writer's book in swarm prevention (78, pp. *40*, *52*).

It will be observed that this method is rather involved because of the need for avoiding loss through swarming; whilst it is much more satisfactory to follow Method 3 given below, this one at least enables the one-hive beekeeper to utilise his queen-cells for increase without the risk of losing his swarm. It is not recommended for the owner of several hives for whom the following procedure is preferable.

Method 2.

(*a*) Proceed as directed in Method 1 (*a*) (p. 161).

(*b*) At the same time form queenless nuclei from *other* stocks which have not made any preparations for swarming. If these nuclei are to remain in the same apiary they should be formed as described on p. 283. If they can be removed to a distance of a mile or more it is better to form them as described on p. 284. They should remain queenless for at least two full days before queen-cells, if unprotected, are given to them.

(*c*) Two days before the young queens of the parent stock are due to emerge from their cells proceed as follows:—

(1) Examine each of the prepared nuclei. If any of them have begun to construct emergency queen-cells (p. 176) these should be destroyed. The nuclei are now ready to receive queen-cells from the cell-raising stock.

(2) Examine the parent (cell-raising) stock and note the positions and number of the best queen-cells. If the number of nuclei is limited to two or three it will usually be possible to select as many combs, each carrying one good queen-cell, for transference to the nuclei. All other queen-cells must be destroyed.

Remove the combs with queen-cells and clear them of bees with a little smoke and a feather. These combs must not be shaken (p. 194). Now insert one comb into the middle of each nucleus, taking care not to crush the queen-cell. Any combs removed from the nuclei should be first shaken clear of bees and then put into the parent stocks. Each nucleus should stand near its parent stock.

If preferred, the queen-cells may be cut out and given to the nuclei. The advantages to transferring the combs are that the nuclei are strengthened and the queen-cells are well received.

As the bees of the nuclei have not developed the swarming impulse it is not essential to abstract the flying bees before the young queens emerge.

(3) Five or six days later remove the nuclei to their permanent locations—this to prevent mating swarms. If selected drones are to be added to the nuclei this is the time to do it (p. 308).

The parent stock will continue its preparations for swarming. This may be prevented as described above, but it is much better if Method 3 described below be followed from the beginning.

Method 3.

By far the best method to follow, if we wish to raise some of the young queens and at the same time to prevent the threatened swarm, is that described in the writer's "Swarming, Its Control and Prevention" p. 40. In this the older bees are separated from the queen, young bees, and brood. Within seven days the young bees tear down the queen-cells and have no desire to swarm. During the same period the older bees, having no queen-cells or means to raise any, also lose the swarming impulse. The queen is then restored to them. The queen-cells usually remain unharmed for three or four days during which time some of them may be used in nuclei formed with brood and young bees from their own stock, or they may be cut out and distributed to nuclei formed from other stocks which have not developed the swarming impulse. Some queen-cells should be left with the queen and the residue of the stock to provide evidence, by their destruction, that the swarming impulse has been lost (78, p. *40*).

Method 4.

Herrod-Hempsall (33, I, p. *712*) describes a method of obtaining 10 nuclei, one with the old queen and each of the remaining nine with one queen-cell, from a stock which is preparing to swarm. Only a very strong stock should be used for this purpose.

When the queen-cells are ready for sealing they are distributed so that each comb of the brood nest carries one only. The comb on which the queen is found is temporarily put aside, e.g., into a nucleus hive. Each of the remaining combs with adhering bees is placed in a nucleus hive and with it two other combs containing stores, but without bees, obtained from other hives. The comb with the queen is then restored to the original hive and with it are placed two combs containing stores, and other empty ones.

His description may be supplemented by the following observations:—

(1) The old-queen nucleus will receive flying bees and will continue preparations for swarming.

(2) The remaining nuclei should not be allowed to lose flying bees until the young queens have emerged. They may be confined to their hives for three or four days by perforated zinc over the entrance, allowing for ample ventilation, and kept in a completely shaded place. When liberated they should be placed in their permanent positions preferably at some distance from that of the parent stock.

(3) Since these nuclei are too weak to be caused to lose flying bees some loss by absconding when mating flights are taken must be expected.

(4) Nuclei so made must be kept warm and provided with ample food. They should be strengthened from time to time by emerging brood from other stocks.

B. *After the swarm has issued.*

Still bearing in mind that we must prevent the issue of second and later swarms we have to consider two conditions:—

(*a*) When the swarm has come out to time, that is on the day after the sealing of the oldest queen-cells.

(*b*) When the swarm has been delayed by bad weather.

Condition (*a*) is indicated by the absence of bad weather preceding the issue of the swarm, by the presence of both sealed and unsealed queen-cells, and by the fresh appearance of the sealed queen-cells.

In condition (*b*) antecedent bad weather will suggest an inspection of the queen-cells. All of them will probably be sealed and the ripest may have the appearance described on page 39.

In case (*a*) we desire to leave the flying bees with the queen-cells until these are more mature. In case (*b*) we should make nuclei at once.

We proceed as follows:—

Case (*a*) which we will call

Method 5.

Hive the swarm in a fresh hive, give it the supers and place it quite near to the parent stock, facing the same way. Five days later move the parent stock to a new position behind, and a few feet away from the swarm. The queen-cells will be mature and flying bees will reinforce the swarm. On the same or the next day divide the parent stock into nuclei. From three to five can be made, or more if the method described on p. 166 be followed. Each nucleus must contain only one queen-cell together with brood and stores. One nucleus may be left in the hive of the parent stock.

The nuclei may be placed near the hived swarm (Pl. XVI) so that when they are finally moved away about twelve days later their flying bees will be received by the swarm. If at this time any of them are not considered to be reasonably populous they may with advantage be reinforced by emerging brood from other hives. In case (*b*) which we will call

Method 6.

Put the swarm into a fresh hive, give it the supers, and place it on the stand of the parent stock. Remove the latter to a new position and proceed to form nuclei at once as described in the preceding paragraph.

CHAPTER IX

UTILISATION OF SUPERSEDURE QUEEN-CELLS

THE natural supersedure of queens takes place much more often than is commonly supposed. It is nature's method of disposing of an old or defective queen and rearing another to take her place.

We do not rely on supersedure for the renewal of queens for the following reasons:—

(1) It may take place at any time, from early spring to late autumn. If in early spring, there are no drones on the wing to mate with the young queen. In late autumn there are very few, if any drones.

(2) It is preceded by a period during which the old queen fails to maintain a good brood nest.

(3) It is succeeded by the discontinuance of brood rearing for two or three weeks before the young queen begins to lay.

(4) There is no outward sign that supersedure is taking place except perhaps the noticeable weakening of the stock.

(5) The young queen may fail to become mated, and if this is not discovered by the beekeeper, the stock will be ruined.

(6) The weakening of a stock by supersedure during late spring or early summer may result in the loss of a crop of honey.

On the other hand supersedure during a nectar flow—when it often happens—results in a diminution of the brood to be fed and consequently a greater storage of honey. During the month succeeding the nectar flow a stock may be so replete with bees that it is not harmed by a temporary cessation of brood-rearing.

Generally then we may say that supersedure may be harmful during spring and autumn but advantageous during late summer.

Supersedure takes place when the queen fails, on account of advanced age or physical defect, to produce a reasonable amount of worker brood. If she is old the amount of worker brood is small and may be interspersed with drone brood. If she is being superseded on account of injury (e.g. the loss of a leg or wing), the brood may be plentiful and normal.

The bees then raise queen-cells—often only one, and seldom more than two or three—which being few in number, are lavishly provided with royal jelly. Even a weak stock whose queen has almost ceased to lay will produce excellent and well-provided queen cells, and consequently it is widely considered that queens reared under the supersedure impulse excel all others. We must not forget however that supersedure is frequent in stocks headed by inferior queens and that the qualities of these are transmissible; we may agree however that in respect of nutrition, which is one of the most important factors in the production of young queens, there are no better conditions in which they can be reared.

It may here be observed that in some forms of imposed orphanage (e.g., where the queen is caged within the hive) the conditions approximate to those of supersedure. The bees are able to communicate with and to feed the caged queen but her inability to lay in the combs probably causes them to regard her as failing and they at once proceed to rear a successor.

When queen cells are raised in the circumstances just described there is no tendency for the bees to swarm, except perhaps when the supersedure takes place during the swarming season, and then only rarely. If therefore we find a single queen-cell, or at the most

PLATE XVII

[Photo: Rev. G. H. Hewison

A SUPERSEDURE QUEEN-CELL

two or three, in a stock which is weak or which has an old or physically defective queen, whether at swarming or at any other time, we should not conclude that the bees are about to swarm. We may safely leave the stock to nature's course or may take advantage of the situation as described below.

If in the course of an autumn examination we find a lively, nervous, young-looking queen where we supposed there was an old one we may conclude that supersedure has taken place at some time during the summer. A careful examination of the combs will reveal the open queen-cell from which she emerged, and our conclusion will then be confirmed.

Should we at any time when drones are available, and particularly in the late spring, find a stock which is superseding its queen, we have at once an excellent means of raising a succession of young queens. If the stock has had a good history we may first make use of its one or more queen-cells as follows:—

Method 7.

Note the date on which the queen-cells are sealed. On the same or next day prepare nuclei to receive the cells in excess of one which is to be left in the stock itself. One or two nuclei will usually be sufficient.

Five days after the sealing of the cells examine the nuclei. Break down all queen-cells they have begun to construct, and give to each a queen-cell from the superseding stock.

Method 8.

If we take away all the queen-cells and give them to nuclei, we find that the stock will continue its attempt to supersede its queen, and fresh queen-cells will be raised. At the appropriate time these can be removed for use in nuclei, and again others will be raised. In this way we may get a succession of excellent queen-cells until the old queen is near exhaustion. We can then leave a queen-cell for supersedure.

If this procedure is followed it will be necessary to maintain the strength of the superseding stock by the occasional addition of a comb of *emerging* brood taken from another stock, and to keep it well provided with food.

In this connection it is interesting to read Doolittle's account of his discovery of this method (23, p. *19*):—

"As soon as I found the cells they were sealed over, and not knowing when they would hatch I at once cut them out and gave them to nuclei. In a few days I looked in the hive again, when I found more cells started which were again cut off and given to nuclei just before it was time for them to hatch. In this way I kept the bees from their desired object for some two months, or until I saw that the old queen was not going to live much longer, when I left one of the cells which they had under headway to mature. By this plan I got about sixty as fine queens as I ever reared and laid the foundation of my present plan of securing queens."

We must not all expect to be as successful as Doolittle. As previously stated a superseding stock may raise only one queen-cell at a time. Seldom does the number of cells exceed two or three. It is for this reason that they are abundantly provided with royal jelly. If however the stock is allowed to become weak, or impoverished, the later queen-cells may be inferior and their larvae indifferently fed.

If the superseding stock is not one whose qualities we desire to perpetuate we may vary the procedure indicated for Methods 7 and 8 by associating with it the method of Substitution thereby combining the advantages of natural breeding and artificial selection. We proceed as follows:—

Method 9.

If the queen-cell or cells be sealed or nearly sealed when first noticed allow them to mature. Five days after sealing transfer them to prepared nuclei. Watch carefully the new batch of queen-cells which will be raised. When these contain larvae of two or three days of age substitute for them larvae one day old, taken from the best stock, in the manner described on page 190. Distribute these cells to nuclei in due course as already described.

In some systems of working for honey production, e.g., the Demaree Method (78, p. *59*) it is usual to raise the brood above the excluder, leaving the queen below. Queen-cells may or may not be made amongst the brood. The bee-keeper usually destroys them. As however they are made under the impulse of what we may call induced supersedure, are few in number, and are built without haste in the presence of the full force of bees and abundant stores, these queen-cells are generally good. They may with advantage be used in nuclei provided that the characteristics of the stock producing them are desirable.

The deliberate use of this principle for queen-rearing is described on p. 202.

CHAPTER X

CONDITIONS UNDER WHICH BEES RAISE EMERGENCY QUEEN-CELLS

MOST methods of artificial queen-rearing depend on the reduction of a stock of bees to a state of orphanage. Deprived of their queen they are soon in the mood and physical condition to raise new queens.

If the queen be removed from a stock the bees are at first unaware of their loss. After a short time however, usually less than fifteen minutes, they begin to show signs that they have missed her. The nurse-bees in the brood nest, whose duty it is to feed the queen with secreted food, probably miss her first. They communicate their sense of loss to the other bees, and soon a state of commotion spreads through the hive. The bees run excitedly to and fro over the combs, frequently stopping to challenge one another and then renewing the search. Not finding her on the combs many of the bees search excitedly outside the hive entrance, running over the alighting board and up and down the hive front. This state of agitation is accompanied by a general humming which, usually within an hour, may so develop in intensity as to be audible at some distance from the hive. Whilst returning foragers enter the hive as though nothing has happened the departure of the outgoing ones gradually diminishes.

It is important that the beekeeper should recognise the signs following the loss of a queen,—the concerted humming and the agitated search over alighting board and hive-front—for at times queens are accidentally and unwittingly killed in the course of manipulations.

The intensity of commotion after dequeening varies in different stocks and in different parts of the season.

The loss of the queen is most acutely felt when she is in full lay and general activity is greatest. The agitation at the entrance may be partly masked during the daytime by crowds of incoming foragers and is therefore more observable in the evening, when foraging has ceased. In some cases only feeble re-action may be noticed, whilst in a few cases,—perhaps one in a hundred—a stock may show no external signs of having lost its queen.

The excitement following sudden orphanage reaches a maximum in about six hours. It persists, in diminishing intensity, until well into the second day. At some time between six and twelve hours however an instinctive communal decision is made that a new queen must be raised and about twelve hours after the loss of the queen the bees begin to construct a number of emergency queen cells around newly hatched larvae. They usually select larvae of the first day, i.e., of the fourth day after the laying of the eggs, but often, in their distress and anxiety to remedy their loss, they construct some of the queen-cells over older larvae, which, as is explained in Chapter IV, is highly undesirable.

The state of orphanage commonly utilised to induce the construction of queen-cells in artificial queen-rearing may be complete or partial. It is complete if the queen is entirely removed, but partial if she is separated from the whole or part of the brood nest by means of a screen which she is unable to traverse but which does not prevent the bees on the other side from being aware of her presence or even passing through to her. The tendency of the orphaned bees to raise new queens depends on the completeness of the separation. Thus, if the queen be placed in a cage covered with small-meshed wire cloth or perforated zinc, and the cage placed cloth-downwards over the combs, the separation, whilst not as effective as the entire removal of the queen, will cause the bees to raise a few—often very few—queen-cells.

The same result is obtained if the queen is confined to one part of the hive by means of a wire-cloth division board whilst the main body of the bees with the young brood occupy the other part.

The confinement of the queen with a comb of brood, bees, and stores in a "Whyte" Cage (p. 190), which is placed in the brood-nest, also induces the building of queen-cells on combs outside the cage, but in this case the number of cells raised is usually very small. The caged queen depends for her food on the bees which are caged with her. The external bees tend to ignore her from the time when they begin to raise new queens.

It is well known that if young brood is taken from a stock and placed in a super over an excluder the bees above the excluder may or may not raise queen-cells on the young brood. As a rule they will not do so but if the raised brood is separated from the original brood nest and queen by an excluder and an interposed honey super, or better still, by two or more supers, queen-cells will be built. The more remote the raised brood is from the queen the greater the tendency of the bees to feel the loss of their queen and to rear others. The writer has found that bees can be made to feel even more separated from the queen if two excluders are used,—one under and the other over the intervening supers. This is an important point and has a bearing on the raising and care of queen-cells over a honey-storing stock without the removal of its queen.

Bees in a state of orphanage will rear more young queens in times of prosperity, that is, when temperatures are high and income of food plentiful, than they will in cold weather or in times of dearth. They will rear more also in a state of complete rather than of partial orphanage. Occasionally however, in continued cold or wet weather, when there is no income of water, nectar, or pollen, and little reserve of stores, a dequeened stock may decline to raise queen-cells.

When a stock is suddenly deprived of its queen it

becomes disorganised for several hours. When their loss is fully realised the bees concentrate on the rearing of a successor, and in their haste to do this they often build some of their queen-cells over larvae of the second or even third day,—larvae which have received some feeding as destined to become workers. The queens resulting from these will be inferior (p. 180), and as they will be the earliest to emerge from their cells they will destroy their younger and better sisters, and one of them will ultimately reign as queen. Without the intervention of the beekeeper therefore, there is a serious inherent disadvantage in rearing queens by imposed complete orphanage.

The complete removal of the queen results in the gradual cessation of brood rearing and consequent retardation and weakening of the stock, with, perhaps, the loss of a honey surplus. This disadvantage however is off-set, in respect of queen-rearing, by the fact that the demands of the young brood on the nurse-bees begins to diminish just at the time when the latter have to make the great effort involved in providing quantities of royal jelly for the young queens.

In the case of partial orphanage however the presence of the queen on the other side of a screen or excluder greatly reduces disorganisation and the bees go more calmly and deliberately about their task. They almost invariably select young larvae of the first day with the result that the chance of the production of hastily-reared imperfect queens is greatly diminished, or is absent.

Some authorities hold the view that queens reared under the stress of orphanage cannot be as good as those reared under the swarming impulse. Whilst we must recognise the ideal conditions at swarming time,—the increasing population, the income of nectar and pollen, favourable temperatures, and the natural tendency to increase, we may maintain that only physical conditions are essential for the rearing of perfect queens and that we can induce these conditions artificially. After all, it is the appropriate nourishment of the young queen larva that is essential, and experience has shown that the finest queens may be reared in the absence of what we may call the emotional conditions associated with the swarming impulse.

The ideal conditions requisite in a stock engaged in queen-raising are discussed in Chapter XIII. Suffice it here to say that they cannot be maintained in a weak colony. For this reason the not infrequent practice of dequeening a nucleus colony in order to cause it to raise queen-cells cannot be too strongly deprecated.

To conclude we may, with reservations relating to exceptions which occur in most matters relating to bees, generalise as follows:—

(1) In complete orphanage many queen-cells are reared, some good, some bad.
(2) In partial orphanage fewer queen-cells are built, all or most of them being good.
(3) The best queens reared in a state of orphanage are as good as any reared under the swarming impulse.
(4) Imposed orphanage gives the queen-rearer the advantage of choice of time and increased facilities for selection.
(5) Only strong stocks, in conditions of prosperity, should be used for the rearing of queens.

CHAPTER XI

REARING QUEENS BY DEQUEENING AND STOCK DIVISION

Simple Dequeening.

Method 10.

It is not uncommon for some beekeepers—having little time or skill, but realising that they must have young queens—to raise them by the simple expedient of removing the old queen from a stock at the end of the honey season and leaving the rest to nature. This course is objectionable for the following reasons:—

(1) There is no selection.
(2) The first queen to emerge from her cell will probably be the worst and will be the survivor (p. 42).
(3) If the survivor fails to mate the stock will be ruined unless a fresh queen be obtained.
(4) Even if the surviving queen mates the stock will be without brood for approximately a month and will be weakened.

The same circumstances arise when a queen meets an untimely death from disease, or by accident,—e.g., is crushed during manipulation; is killed by robbing bees; or is balled by her own bees through unseasonable disturbance. As the bees are not guided by reasoning but by instinct they become so eager to repair their sudden loss that they often select larvae of different ages, including some of two or three days old, from which to raise queens. A three-day old larva will have been fed as a worker and the development of her ovaries will have been restricted for two days (p. 65). It can be fed on royal jelly for a further two days only, that is for only half the time necessary to produce a normally developed queen. The resulting queen will be small, of limited fertility, of short useful life, and may later give trouble by passing through queen excluders.

Remembering that a larva—whether queen or worker—is sealed over on the fifth day after the hatching of the egg we reason that queen larvae sealed on the fifth day were selected to become queens at or about the time of hatching and we assume that they will have been fully fed as such from their first larval day. Larvae selected at one day old will be sealed on the fourth day; at two days old on the third day; and at three days old on the second day. If therefore we examine a dequeened stock on the fourth day after de-queening and destroy all queen cells which are then sealed we are reasonably sure that the remaining (uncompleted) queen-cells will have received the royal food during the full natural period and that the queens will be as good as possible in the circumstances.

The practice of raising a new queen by simple dequeening is therefore not wholly to be condemned. We can make the best of it by:—

(1) Reserving the old queen in a nucleus (p. 161) so that she can be restored for a time in case of the loss of the young one.
(2) Destroying all *sealed* queen-cells on the fourth day and leaving one or more open ones best provided with royal jelly.
(3) Feeding throughout the five days.

When the young queen has become mated, as shown by the presence of sealed worker brood, the old queen may be destroyed and her nucleus added to the stock by the newspaper method.

Method 11.

An improvement of the foregoing method is effected as follows:—

Remove the queen and all the combs containing eggs and young brood, together with the bees covering them, and put them with some additional shaken bees into a nucleus hive. Place this nucleus on a new stand. Between the remaining brood combs insert a comb of eggs, taken from the best stock in the apiary and secured as described on page 194. In other respects proceed as advised for the preceding method. As it is difficult to make sure that all eggs other than those introduced have been removed it is important to examine for undesired queen-cells on the fourth day.

An alternative course, which many will prefer, is to make a nucleus with the old queen within the hive and so to cause the remainder of the bees to raise a new queen. The procedure is as follows:—

Method 12.

We assume that the honey has recently been removed and that therefore the stock is strong in bees occupying the usual ten or twelve combs in one brood-box. For convenience we assume also that the combs are parallel to the entrance and that the hive is "single-walled". Modifications for other types of hives are referred to below.

By means of a close-fitting division-board, or one which fits into grooves, divide the hive into two compartments, the rear one to contain three combs and the front one six or seven. This division-board must fit closely to the floor and sides of the hive and must be packed at the lugs so as to make it impossible for a bee to pass from one compartment to the other. It may have a small perforated-zinc or wire-cloth panel if desired, but this is not a necessity.

A small entrance for the rear compartment may be contrived in the floor of the hive, or a $\frac{3}{4}''$ hole, filled by a cork when not in use, bored through the rear hive-wall. The bees in the larger compartment will use the ordinary hive entrance, —opened to full width.

Cut a small quilt, of canvas or American cloth, of a size to cover the smaller compartment, and fix one long edge of it by means of drawing pins to the top edge of the division-board. Cut a similar larger quilt to cover the larger compartment. This need not be fixed to the division-board. Either compartment can then be examined without disturbing the other.

Into the smaller compartment place:—

(1) A comb of brood with queen and adhering bees.
(2) Extra bees shaken from a well-covered comb,—this to compensate for bees which will return to their customary entrance.
(3) A comb of stores with adhering bees.
(4) A third comb with adhering bees but little or no brood or stores.

The remaining combs, with bees, will occupy the larger compartment. Amongst them should be the comb in which the queen last deposited her eggs, or preferably, a comb of eggs from a better stock. This should be placed in the middle of the compartment and given rather more space than it previously occupied so that there will be ample room for the construction of queen-cells.

The flying bees of the old queen's compartment will return to the queen-raising stock in the front compartment. This should be fed for at least five days whether the combs are well stored or not. The queenless bees will raise a number of queen-cells on the comb of eggs.

On the 4th day examine the queen-cells. Some may be sealed whilst others will be open. Destroy the sealed ones and leave one or more of the open ones which appear to be the best supplied with royal jelly. The first young queen should emerge from her cell on the 12th or 13th day. If there are no accidents and the weather is favourable she should begin laying at some time between the 21st and 36th days. Should she not be laying during this period remove her and insert another comb with eggs.

In early autumn both compartments should be well stocked with brood and ready for uniting. Proceed as follows:—

(1) Remove the old queen.
(2) One or two days later, at evening, raise both quilts at one end of the division-board, place across this a piece of batten or round stick about $\frac{3}{4}''$ in diameter, and replace the quilts on it. On each side of the stick there will be a bee passage through which the bees will slowly pass and unite peaceably. A day or two later the division-board may be withdrawn.

This is a satisfactory method of requeening for the beekeeper who has only one or two hives and who does not wish to practise artificial methods. It leaves the stock stronger than it would normally be, well provided with brood which will produce the young bees necessary for good wintering, and headed by a young queen bred in reasonably good conditions.

In the event of the young queen being lost

or becoming a drone-breeder, the front compartment, when definitely known to be queenless, must be united to the smaller compartment which retains the old queen until a successor can be obtained and introduced. If later than the middle of August it is not desirable that the stock should be allowed to make a second attempt to requeen itself.

In most hives the brood boxes can be placed so that the combs are parallel to the entrance. In cases where they must be at right angles to it the entrance for the small compartment is bored through the hive side, and the portion of the main entrance between this side and the division board completely closed.

For W.B.C. and other double-walled hives the small entrance can be provided by a hole bored in the floor near the back of the hive with a rimmed alighting board half-an-inch beneath it.

Method 13.

Young queens are produced by imposed partial orphanage in the working of the writer's system of Swarm Prevention (Pl. XVIII). The principle of this can be given briefly here. For full details the reader is referred to the book on the subject.

The queen with all the field bees are at first partially and then completely separated from the young bees just before swarming time. The former build a new brood nest and continue to work in the supers, whilst the latter raise queen-cells one or more of which are allowed to produce young queens. At intervals the working stock is reinforced by the bees which reach the flying stage in the queen raising stock.

In good weather and with ample food and water the young bees rear good queens. In adverse circumstances however, inferior queens, or even none at all, may be the result. In order to secure consistently good results the following addendum has been inserted in the 7th Edition of the book (78, p. *97*). It is quoted here in full for the benefit of readers who may be using any of the earlier editions:—

Addendum to pp. 18–35 of "Swarming, Its Control and Prevention". 7th Edition.

"This method was devised for the prevention of swarming. Its use for queen rearing is incidental and secondary. In cold weather with lack of income the bees in Box A may construct poor queen-cells or may refrain altogether from making them. Experience has shown that the best queens are raised when bees of all ages are present, and income, especially of water, is continuous.

"Mr. J. F. Bramwell, of Exmouth, has given much attention to the quality of queens raised. By modification of the procedure in the case of one stock he secures from it all the queen-cells he needs and these are raised under ideal conditions. He proceeds as follows:—

"As far as possible equalise the stocks during May by helping the weaker ones with brood so that they may all be ready for Method 1 at the same time. On the date chosen for the application of Method 1 put a frame of unwired foundation into the centre of the brood nest of the best stock. Apply Method 1 to all other stocks. Six days later select one of the treated stocks for queen rearing. Break down all its queen-cells in Box A. Remove two or three combs and substitute for them:—

"(1) A division-board feeder filled with diluted honey or syrup: (2) The frame of foundation, now drawn out and containing eggs, from the

PLATE XVIII

[*Photo: L.E.S.*

TRIPLE QUEEN-REARING NUCLEI OVER SWARM-CONTROL BOARDS

best stock. (This comb may be divided into two portions by cutting horizontally through it so as to leave scolloped edges. The lower part may be fixed into another frame, scolloped edges downwards. Easily removable queen-cells will be built along these edges) (Pl. XIX).

"Now temporarily remove Box B to another stand and put Box A in its place under the supers. The stock will become very strong and will raise queen-cells. When these are sealed (six or seven days later) distribute them to nuclei—deprived of their queen-cells—made from the several Boxes A, or to Boxes A themselves after first destroying their queen-cells. For greater security each queen-cell may be suspended in a spiral cell protector, (p. 295). Only one queen-cell should be allowed in each nucleus and in Box A of the cell producing hive. If weather conditions are good, Boxes A and B of this hive may resume their normal positions, i.e. Box A above the screen-board and Box B below the excluder."

As previously stated partial orphanage induces bees to raise queen-cells. The two following methods are based on this principle.

Method 14.

Place the queen and about a dozen of her bees in a well-provisioned queen-cage. Suspend this cage between combs or place it,—wire-cloth downwards—over the feed-hole in the quilt. When the bees find that the queen cannot reach the brood they consider themselves queenless and construct queen-cells, generally only a small number. Sealed cells should be destroyed, as in some preceding methods, on the fourth day. The remaining ones may be removed on the tenth day for use in nuclei and the queen liberated. Alternately, one queen-cell may be left for the

PLATE XIX

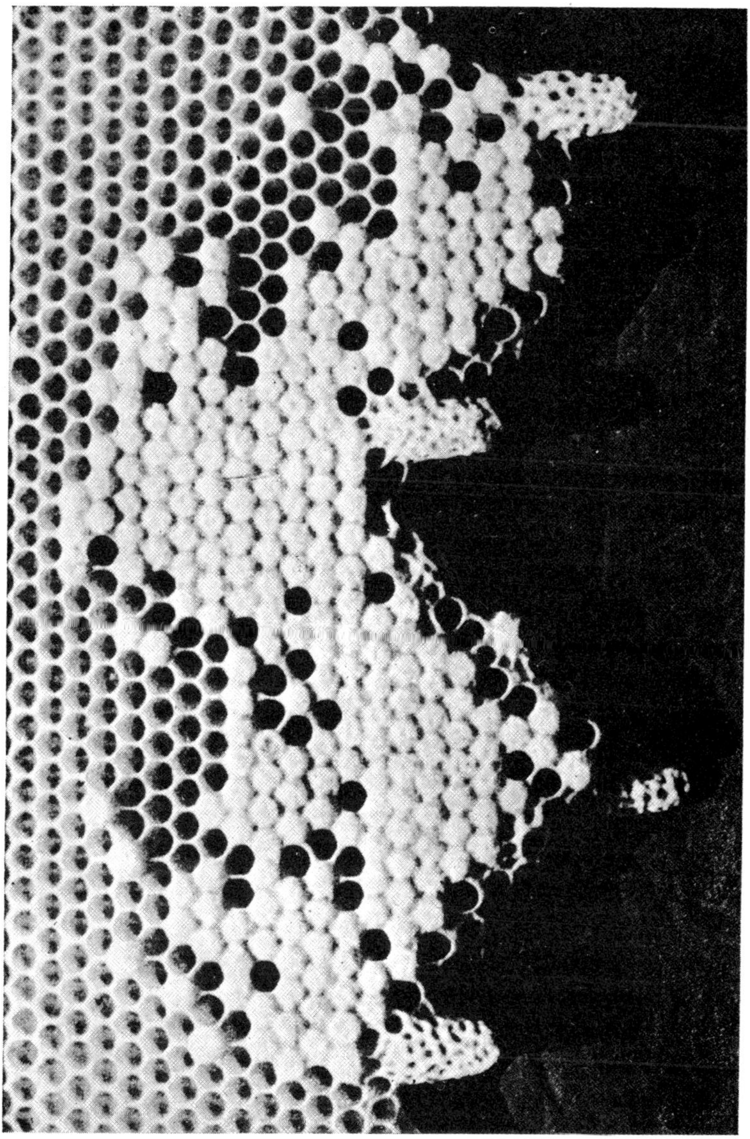

[Photo: J. F. Bramwell

QUEEN-CELLS ON SCOLLOPED COMB. (SWARM PREVENTION METHOD)

requeening of the stock and the old queen removed. A disadvantage of this simple method is that whilst a queen and her attendants will live without harm for a long time in a well-provisioned cage kept in a suitable temperature and away from the hive, it is otherwise when she is exposed to the bees through wire cloth. The efforts she constantly makes to rejoin her stock seem to wear her out, and if imprisoned until the queen-cells are ready to be taken away she is of little use afterwards and is not welcomed by the bees, which, as the queen-cells mature, begin to neglect her. When the cage food is used up she may starve.

Method 15.

Confinement of queen in a "Whyte" cage.

The "Whyte" cage primarily designed as an introducing cage, may be used to induce bees to raise queen-cells. It is constructed of wire cloth with a sliding metal top, and is of size and shape to contain a brood comb and the necessary bee spaces. The sliding top is provided with a $\frac{3}{4}''$ hole which can be closed or opened by a movable metal cover. It is used as follows:—

The combs are examined and the one with a good supply of stores and little brood is selected. This is put into the cage with a moderate covering of bees. (An excess of bees will use up the stores too quickly.) The queen is run into the cage, which is then closed. The cage is hung in the brood-nest—preferably away from the centre. As the "Whyte" cage occupies more space than a single comb it may be necessary to withdraw two combs temporarily from the hive to make room for it.

The bees outside the cage will now raise queen-cells, but usually only a small number. These will be good. As in the preceding cases they should be inspected to ensure that they are unsealed on the 4th day. The queen-cells must be removed by the 10th day and the queen then liberated by the opening of the metal cover.

The advantage of the use of this cage is that the imprisoned queen is unharmed while her stores last, and is therefore suitable for retention. Although the procedure appears to be easy from a mere reading of it, it is really a little troublesome, and one or two important precautions must be observed:—

(1) The cage should be removed as soon as it is known that the queen has emerged from it. This should be indicated by the presence of eggs in other combs within two or three days,—induced if necessary by feeding. If no eggs are found the queen must be taken from the cage.

(2) There is always the danger that the stores in the cage may be insufficient if the period of imprisonment be prolonged. If this happens the bees external to the cage will not, as might be supposed, feed the queen, and both she and her bees may be found dead in the cage.

Method 16.

Substitution of Larvae.

Any of the foregoing methods may be improved by what is known as the method of substitution. This consists in removing the larvae from open queen-cells and substituting for them larvae of the first day taken from the best stock in the apiary. The queen-cells most suitable for this are those containing larvae of the 2nd or 3rd day which are well supplied with fresh royal jelly.

The original larvae are lifted from their cells with the grafting tool, (p. 237) and under suitable circumstances of temperature etc., the selected larvae are put in their places as described on p. 241. Great care must be taken that the larvae are floated on to the surface of the royal jelly and not drowned in it. They should not be deformed by contact with the tool but should retain the shapes and relative positions they had in their original cells. Each cell as treated should be marked in some way—Herrod-Hempsall recommends the insertion of a match stick into the adjacent comb—and all should be returned to the cell-raising stock without loss of time.

CHAPTER XII

PROVISION OF EGGS OR LARVAE FOR QUEEN-REARING

If we deprive a stock of its queen we find, as a rule, that young queens will be emerging from their cells on the 12th day afterwards. This shows that the bees normally start their queen-cells over eggs of the 4th day, that is when they are hatching. If however we provide queenless and broodless bees with a limited number of eggs of the first or second day, and nothing older, they will construct queen-cells over those that mature first and will often ignore the rest, with the result that the number of queens raised may be disappointingly small.

In preparation for queen-rearing therefore we should take pains to select eggs at the beginning of their 4th day, or better still larvae of the day on which they emerge from eggs in order that the nurse-bees may be employed in queen-raising without delay.

It is desirable also that the queen-cells be formed where they can easily be removed. If a portion of the egg-bearing comb be cut away queen-cells are likely to be built in the space so made rather than on the face of the comb, and if the lower portion of the comb be removed the cells will mostly appear along the lower edge of the remaining part of the comb. If no space is provided the bees build queen-cells on all parts of the comb and often in little groups where they are so close together that it is impossible to separate them without injury to some.

We may do much to encourage the bees to construct their queen-cells where we want them, that is at convenient distances from one another and where they may be easily removed.

To obtain a comb of eggs or first-day larvae it is usual to insert a comb of unwired foundation into the middle of a well occupied brood-nest five or six days before the newly emerged larvae are needed. A shallow frame is usually sufficient and is more easily handled than a large one.

In favourable circumstances of temperature, income, and crowded brood-combs, the foundation may be drawn into comb on the first day and the queen will be likely to deposit eggs in it on the second. Three and a quarter days later, that is late on the fifth day or early on the sixth day the first eggs may be hatching. The comb should therefore be examined on both these days and utilised for queen-rearing whilst the larvae are less than 24 hours old. Older larvae should on no account be present (p. 65). Direct light, and if necessary, a magnifying glass, will assist in determining the ages of both eggs and young larvae, the appearances of which are illustrated in Pl. VIII.

When removing the prepared comb of eggs it is necessary to dislodge the bees from it. This must be done with care, first because the queen will most likely be on this comb, and secondly because violent shaking will disturb both eggs and larvae from their proper positions, and this, whilst not killing them, will cause the bees to reject them for queen-raising. Although a slight vertical shake will dislodge some of the bees, and probably the queen also, without affecting eggs or larvae, it is better to proceed as follows:—

Hold the frame by one of the ends of the top bar in the left hand and allow the opposite bottom corner to rest on the other brood combs. Urged by a little smoke from above the bees will soon begin to run down into the brood-nest. The last few may be removed with a feather (Pl. XXXI).

The cleared comb should not be exposed to direct

heat of the sun, to drying winds, or to rain. It should be carried in a warmed and closed nucleus box (p. 239) to the cell-raising stock, or to the grafting room (Chapter XIV).

Cowan method.

Cowan (20, p. *126*) recommends that the comb containing eggs or young larvae be cut away close up to the cells most suitable for queens, and that the particular cells which it is desired that the bees shall select be enlarged by means of a small rounded piece of wood. The cell walls, especially those on the under side, are broken apart without interference with the egg or larva contained in the cell.

Miller method.

Dr. Miller (62, p. *352*) prepared the comb in which the eggs were to be laid in the following way:—

Into the top bar of the frame triangular strips of foundation are fitted, each about $1\frac{1}{4}''$ wide at the top and coming to a point about $1\frac{1}{4}''$ from the bottom bar. A space of 2″ is left between successive strips. As these strips are delicate they are supported by fine wires which can easily be cut with scissors when queen-cells are removed.

The frame is inserted between two combs of brood in the stock with the breeder queen. All other combs of brood are removed to discourage the construction of drone cells in the prepared frame. At the end of a week the strips will be drawn out and should contain young brood. and eggs. They are trimmed with scissors or knife so as to leave larvae of the desired age at their edges. Most of the queen-cells will be built here.

To obtain a good and convenient comb the breeder stock must not be too strong or heavily fed,—otherwise the whole frame will be rapidly filled with comb and there will be no point in using the strips.

Miller says that if such a prepared comb be placed in a strong queenless cell-raising stock it will provide as many queen-cells as one can desire.

Pechaczek method.

Maisonneuve describes what is known in America as the "Case" method, ascribed originally to Hans Pechaczek, Vienna (62, p. *356*), the main purpose of which is to prevent the bees from building queen-cells so close together that they cannot be separated.

A new comb, or sheet of foundation, is inserted in the centre of the stock from which queens are to be raised. In four days, or later, this will be filled with eggs and young larvae. It is then withdrawn, cleared of bees, taken to a room with a temperature of 25° to 30° C. (i.e., 77° to 86° F.), and laid flat on a table, the side better furnished with larvae of the required age uppermost. By means of a warm knife two adjacent rows of cells running the whole length of the comb are pressed down and so destroyed. The next row of cells is left undamaged. The whole face of the comb is treated in this way, two rows of cells being destroyed for each third one left intact. The rows of intact cells are then dealt with. Of every three consecutive larvae two are removed and one left. The face of the comb then carries cells containing larvae each isolated from neighbouring ones by nearly ½″ so that queen-cells built on them can be cut away separately. Cells over the supporting wires are not preserved and larvae and eggs remaining in the crushed cells are removed with a small stiff brush.

The comb is then given to a cell-raising stock. It is not hung in the hive in the usual way but is placed horizontally over the frames, the prepared face downwards. To allow space for the construction of queen-cells it is raised at least an inch by the insertion of one empty frame exactly beneath it. In hot weather a central wooden support will be needed to prevent the comb from sagging. Quilts are laid lightly over the upper surface and a lift is provided to accommodate them. Maisonneuve points out that the cell bases (midrib) must not be broken or the cells on the opposite face of the comb injured, and recommends preliminary practice on an empty comb.

It will be realised that there are some disadvantages in this method. The procedure is delicate and tedious. A large proportion of the cells will be rejected and some of them will be joined by brace-comb. It is better to separate the selected cells by three rather than two worker-cells.

The Pechaczek and Miller methods of securing eggs or larvae for breeding have the advantage that a larger number of queen-cells is raised on the prepared comb given to queenless bees than when a plain comb of eggs is used. For the small scale queen-breeder they are not worth the extra work involved and they have no advantage for the large scale breeder who practises the better methods described in Chapters XIV to XVII.

The transference of young larvae from their original cells to artificial ones, as used in the methods described in Chapter XIV is attended by risks which often result in failure. The tiny, delicate creatures may be chilled, over-heated, dried, starved, drowned in royal jelly, improperly deposited in their new cradles, or wounded by the transferring instrument. To avoid these risks Henry Alley (7, p. *90*) in 1883 described a simple and successful method of transferring them to desired positions without disturbing them in their cells.

Alley method.

Having first obtained a comb of eggs and young larvae from his breeding stock he laid it on a flat surface and with a sharp warmed knife or razor cut a strip of comb containing a single line of occupied cells on the upper face of the comb. He then cut away the cell walls almost to the midrib leaving the cells on the under side of the comb undisturbed. Of the larvae in the shortened cells he removed two out of every three, thus leaving every third cell occupied, and at a distance of nearly ½″ from the next.

Having previously cut away the lower half of an old and empty comb so that the lower edge of the remaining half was a convex curve, he attached the prepared strip of comb to it by means of a small brush and some melted wax containing a little resin.

The cells on the undamaged side of the strip were so attached to the comb, that the shortened cells containing the selected eggs or larvae hung downwards. The convexity of the comb edge and the spacing of the larvae along it facilitated the removal of the queen-cells when mature.

Benton tells us that J. M. Brooks had previously used this technique but had fixed his strips "to horizontal bars nailed into ordinary comb frames",—the equivalent of modern cell-bars (p. 233).

The prepared comb or frame was then to be given to a cell-starting stock (Chapter XIII).

The main points to be observed in the provision of young brood from which to rear queens are:—

(1) A comb of foundation is inserted into the middle of the brood-nest of a prosperous stock headed by a queen of approved qualities.
(2) The comb should be inserted at least 5 days before the cell-raising stock is formed. The exact time for the latter operation can be fixed only when hatching eggs are found in the prepared comb. If honey and pollen are coming in this should be on the 5th or 6th day. If there is a dearth of nectar or if the weather is cold, the proper time may be later, but these circumstances can be corrected by feeding.
(3) Larvae approximately 24 hours old should be presented to the cell-raising stock.

(4) If eggs or young larvae are to be used without removing them from their original cells the comb should be so treated that the resulting queen cells are not built close together and are therefore easily removable when mature.

The Breeder Queen.

The queen selected for breeding should be of pure race if possible (p. 147) and at least one year old. At that age there will be some evidence of the qualities of her progeny, e.g., disposition, prolificacy, industry and productivity, tendency to swarm, and ability to winter well. Apart from the rare chances of mutation (p. 131), her genetic constitution does not change, so that she may be used as a breeder as long as she continues to lay fertilised eggs. If for example she lives through a fourth season that is a qualification in her favour for her progeny may also be long-lived (pp. 136–137).

Accessibility of breeder queen's brood nest.

It is an advantage if the stock containing the breeder queen be limited to one brood-box, or to a large nucleus hive. The frequent removal of excluder and supers is then avoided and the risks of injuring the breeding queen diminished. The writer keeps his breeding queen in a full stock but she is confined to a single brood-box over the excluder, the supers being underneath the excluder. With this arrangement larvae for breeding are obtained quickly and with a minimum of disturbance.

CHAPTER XIII

INDUCING BEES TO START QUEEN-CELLS

We may now consider some of the more approved methods of raising queen-cells and the steps necessary to secure that they are started in the best conditions.

In some cases they are started and completed in the same stock. In other cases, especially when many are needed, they are started in one stock and completed in another. In respect of the latter cases we shall be mainly concerned in this and the ensuing chapter with the starting of the queen-cells, leaving the consideration of their completion to Chapter XV.

The queen-cells may be commenced by,—

A. Bees in possession of their queen and brood.
B. Queenless and broodless bees in confinement.
C. Unconfined queenless bees with or without some brood.

A. Bees in possession of their queen and brood.

The late Dr. Miller (51, p. *237*) was perhaps the first to rear queens in stocks having laying queens. He records the following experience:—

"Upon a hive containing a colony had been piled four stories of empty combs for safe keeping. To make sure that the bees would not neglect the care of the most distant combs I put a frame of brood in the upper storey. A few weeks later I found a laying queen in the upper storey, with the old queen still below. The bees that had gone up to that frame of brood were so far from the queen that they had reared a queen of their own. A hole in the upper storey had allowed the flight of the young queen without invading the domains of her mother."

Dr. Miller adds:—

"Neither is it absolutely necessary to have a queen-excluder between the stories. In lieu of an excluder I have used a cloth with room for passage at the corners. Neither excluder nor cloth is absolutely necessary; distance is enough".

It is now commonly believed that bees on brood which cannot be reached by the queen consider that she is failing and rear other queens under the supersedure impulse. Consequently we are able to use powerful queen-right stocks for the completion of queen-cells already begun (Chapter XV).

Illingworth (41) makes use of this principle for the production of young queens for his own apiary. Each of his Langstroth hives carries a super as an extra food chamber all the year round. The queen has access to this super so that the brood chamber is thereby enlarged. At a convenient time the super is utilised for queen-rearing. The queen is confined to the lower brood-box by an excluder which is placed between it and the super. A special shallow frame, with a bar of artificial cells containing grafted larvae (p. 235), is then inserted into the middle of the super. The other shallow combs of the super are later used for forming nuclei for the reception of the queen-cells when mature.

Method 17.

When young brood is raised from a brood-nest to a super above an excluder, the bees may or may not construct queen-cells under the supersedure impulse. If however there are three or more well-occupied supers and the young brood is inserted in the topmost one, queen-cells of good quality will be made and completed. The tendency to raise them is in proportion to the remoteness of the super from the queen and is increased if a second excluder be inserted under this super. The inserted brood should, as always, be taken from a selected stock.

Although this is perhaps the easiest method of obtaining good queen-cells we may be disappointed with their number. More will be obtained from a natural comb containing eggs and young larvae than from artificial cups, and success will depend on the time of year and the strength and prosperity of the stock. When the queen-cells have been removed (p. 42) the procedure may be repeated. We may note here that the construction of such queen-cells in a queen-right stock does not, in itself, lead to swarming.

Method 18.

Beekeepers who practise the Demaree system may utilise it for queen-rearing. The essential feature of this system is that the whole of the brood-nest, except one comb with the queen, is raised above the excluder. The bees above the excluder often raise one or more queen-cells, especially if a super intervenes between them and the excluder. In these circumstances they will usually raise a limited number of queen-cells on an inserted comb of eggs or on grafted queen-cell cups if these are given to them after the removal of their own queen-cells. Failures are likely to occur however on account of the presence of the stock's own young brood, and in any case only a small production of queen-cells may be expected.

Method 19.

Audibert (3, p. 47), in his book "Plus de Miel" describes a method, which he ascribes to Demuth and Root, of causing the bees of a raised box of brood to build queen-cells. The stock, consisting of two boxes of brood is divided so that the queen is in the lower one, and suitable young brood for raising queens in the upper one. The two boxes are separated, not by excluder, but by two super-clearer boards, one placed on the other,—the rims leaving a space between them. The clearer holes are closed.

The lower of the boards is pierced with a 1″ hole towards the front of the hive and the upper one with a similar hole towards the back of the hive.

The bees can then pass freely from one box to the other that is,—

Upwards through the hole in the lower board, across the space between the boards, and through the hole in the upper board; or downwards in reversed directions. In the case of a supered stock an excluder is necessary under the supers, the double board being placed over them. The partial separation of the bees by the boards is sufficient to cause those in the upper brood box to raise or accept queen cells.

Hudson (37, p. *26*) has applied this principle to the writer's Swarm Control method, the advantage being that all the bees have access to the queen-rearing part of the hive and that therefore better queen-cells are built.

Method 20.

The beekeeper who possesses a long hive made to contain 15 to 20 combs may with advantage use it for queen-rearing.

Let us assume that the hive will accommodate 15 combs and a division board. We may proceed as follows:—

Divide the hive into two compartments (one for 10 and the other for 5 combs) by means of a close-fitting three-ply division board sliding in grooves and reaching to the floor. The edges of the division board should be vaselined to make it easily withdrawable.

The central portion of this division board, about 8″ by 5″ is cut out and the space covered by

(*a*) Sheet metal in winter, or
(*b*) Excluder zinc whilst the smaller compartment is being furnished with combs of brood, or
(*c*) Perforated zinc whilst queen-cells are being raised.

It will facilitate working if separate division boards for (*b*) and (*c*) are available.

As the stock (on the 10 combs) develops in spring it is supered in the usual way. When congestion of the brood nest approaches two combs of sealed brood with bees but without the queen are placed in the smaller compartment behind the division board, which is now fitted with excluder (*b*). Foundations or empty combs are substituted for the transferred combs.

About a week later two other combs with sealed brood but without very young brood or eggs, are similarly transferred to the small compartment. If this holds more than 5 combs so much the better, for the process can be repeated.

Care must be taken that the transferred combs have some stores of honey and pollen and that finally a central space is left for the insertion of a prepared comb of eggs and young larvae, or grafted queen-cells. If possible unsealed brood of more than three days of larval age should be placed on each side of this space to attract nurse-bees to it.

The bees of the smaller compartment must be able to pass in and out of the hive only through the excluder of the division-board.

The successive transference of brood will tend to delay swarming. If queen-cells are built over eggs or young brood inadvertently placed in the smaller (queen-rearing) compartment they must be destroyed.

The stock is now ready for queen-raising. Division board (*c*) is substituted for (*b*) and the prepared comb of eggs or young larvae is inserted in the central space of the smaller compartment. The confined bees start queen-cells and twenty-four hours later division board (*b*) is substituted for (*c*). All the bees of the hive then have access to the queen-cells which are completed in due course.

The queen-raising compartment should be fed with honey and water whilst cells are being built.

By this method successive batches of queen-cells may be raised without disturbance of nectar-gathering.

B. The starting of Queen-cells by confined queenless and broodless bees.

The principle involved is that well-fed bees, if kept queenless and broodless for a few hours, will readily accept and rear queens from hatching eggs or young larvae presented to them, whether in combs or artificial queen-cells.

As this is generally recognised as an effective method of inducing bees to raise queen-cells it is desirable to consider in some detail the necessary conditions of the bees, and the construction and use of the swarm-box which is used.

Method 21.

The Bees.

To obtain the necessary bees we must disturb a strong stock, for the greater the number of bees used the greater the number of queen-cells likely to be built.

To ensure the maximum of success we observe the following requisites:—

(1) The bees used must be taken from a strong stock, preferably from one which is increasing rather than decreasing in strength. May and June are therefore the best months and July and August less favourable. Methods of building up a stock to great Strength are given on pages 249–253.

(2) Bees of all ages should be present, but there should be a large proportion of nurse-bees, i.e., those taken from brood combs.

(3) Pollen, honey, and water must be present, but no eggs or young larvae.

(4) The stock from which the bees are taken should have been fed with syrup—or if occupying supers with little income, with dilute honey—for three or four days previously. The nurse-bees will then be in good condition for the secretion of royal jelly.

(5) Few drones should be included. It is better if there are none, for drones receive brood food from the worker bees, and if they are numerous the nurse-bees will have insufficient secreted food to provide royal jelly for the queen-cells.

This is an important point which appears to have been overlooked by other writers. A year ago the writer used bees from a strong colony kept for rearing selected drones. They were given sixteen grafted queen-cell cups, but only two were accepted and these were small and useless. The large proportion of drones had evidently exhausted the resources of the nurses.

(6) Bees about to swarm are excellent for cell-raising but if they are used the problem of controlling the swarming provides an additional care. Confinement in the swarm box together with the raising of the queen-cells does not diminish the tendency to swarm.

(7) Bees which are confined for some hours need a supply of water. This can be provided either by a comb into which water has been sprayed until the cells are about half-full, or by means of a division board feeder, as sold by some appliance dealers, which is filled with thin sugar syrup.

The swarm-box.

The box (Pl. XX) in which the queenless bees are to be confined for about 24 hours is similar in construction to an ordinary brood box but of only sufficient width to take either three or five standard brood frames,—the latter size when many queen-cells are desired.

The interior width of the smaller box should be 5″ and of the larger one 8″. The depth should be sufficient to allow a space of $1\frac{1}{2}''$ beneath the frames. The floor consists of a sheet of perforated zinc or coarse wire cloth tacked to the underside of the box and kept from the ground by two wooden cleats nailed under the ends. This will afford sufficient ventilation for the confined bees, but in hot weather additional ventilation may be provided by means of two 1″ holes bored through the lid of the box, one towards each side, and covered by perforated zinc. When not needed for ventilation the holes are closed by corks, or covered by feeders.

A shallow, easy fitting cover is provided as illustrated in Pl. XX. As it is undesirable to remove this after the box has been furnished with bees arrangements must be made for a portion of it only to be easily removed when a frame containing eggs, larvae, or queen-cells is to be inserted or withdrawn. For the smaller (3-comb) box the central portion of the cover, 2 in. wide and of the whole length of the cover, is left open. The resulting space is filled by a strip of wood of the same width but slightly longer. Its long edges, and their seating, are bevelled so that the strip drops into place and makes a good fit. The strip is cut in half to form two slides which can be easily withdrawn or replaced under a carbolic cloth (p. 272). These can be seen in Plate XX.

Through the fixed portions of the cover two 1″ holes are bored to serve for feeding or, when covered by perforated zinc, for extra ventilation.

In the case of the 5-comb box two spaces, one on each side of the position of the middle comb, may be provided, thus permitting the insertion of two combs of eggs or newly hatched larvae, or two sets of grafted artificial queen-cells. The writer finds however that the larger swarm-box fitted for the reception only one set of 20 grafted cells is the more satisfactory.

The stocking of the swarm-box is easy especially if a funnel is used (p. 214). Once it is furnished with bees it can be used for the production of a second and third batch of queen-cells. Since

PLATE XX

[*Photo. L.E.S.*

SWARM-BOXES

however each batch of acceptances will be less than the preceding one it is more satisfactory to use a fresh lot of bees each time (p. 84).

Preliminary preparations.

1. Five or six days before it is intended to commence queen-rearing place a frame containing a sheet of worker foundation in the centre of the brood-nest of the stock from which it is desired to breed. By the fifth day this should be drawn out and filled with eggs. If some of these are hatching the larvae will not be more than one day old and this will be the favourable moment for using them. If none are hatching wait until the next day.

2. Prepare the swarm-box as follows:—

   (*a*) 3-*comb size.* Insert one comb well provided with honey and pollen, and another comb containing water or sweetened water, or instead, a division board feeder filled with thin syrup. The cappings of the honey cells should be bruised. Leave a central space for the insertion of the comb or frame to be used for queen-cells.

   (*b*) 5-*comb size.* Insert two combs of honey and pollen, and one containing some water (or preferably a division board feeder filled with thin syrup), leaving two spaces, one on each side of the middle comb which should be one of those containing stores. Alternatively, insert a third comb of stores leaving only one spade for the frame to be used for cell-raising. Bruise some of the honey cappings.

Stocking the swarm-box.

The following are directions for stocking a swarm-box:—

(1) Select a strong stock, preferably one working well in supers and containing few drones.

(2) At about 9 or 10 o'clock in the morning remove the supers and place them on one side.

(3) Find the queen and place her with the comb on which she is found temporarily in a nucleus hive.

(4) Into a 3-comb swarm-box quickly shake the bees from five or six brood combs. Into a 5-comb swarm-box shake the bees from all the combs of the brood chamber. Do not waste time in trying to dislodge the few bees which will cling to the combs after the first shaking. Some bees will escape but they will be the older ones and the less valuable for our purpose. This operation is simplified by the use of the funnel described below.

(5) Place the closed cover on the box at once and remove the whole to a cool shady place, e.g., a cellar or the basement room of a house, making sure that ventilation is not impeded.

(6) The comb of bees with the queen is now restored to her hive together with the combs of brood from which the bees have been shaken, and the supers are replaced. The bees with the queen and the returning foragers, together with those in the supers will be sufficient to care for the brood until the swarm-box bees are returned.

Use of the swarm-box.

(1) After six or seven hours' confinement the bees, having had no brood to feed and

being replete with food, will have accumulated much brood food and will be ready to feed the young larvae to be given to them.

(2) When the bees have been confined for about five hours, "graft" the prepared artificial queen-cells and insert the frame carrying them into the swarm-box as described in detail in Chapter XIV.

Jay Smith (77, p. *34*) finds that confinement for two hours is sufficient to cause the bees to start queen-cells but the writer considers that a longer period is desirable to ensure an increased accumulation of secreted brood-food in the nurse bees.

(3) Take the swarm-box into a room (e.g., the honey house) with moderate temperature. Place a feeder containing honey and water in equal proportions, or syrup made of equal volumes of sugar and water, over the feed hole. If there is a second feed hole in the cover, and if honey extracting is in progress, place a piece of queen-excluder over the hole and over it invert a basin of cappings fresh from the combs and dripping with honey. Nothing is so stimulating as this for the bees carry off the wax as well as the honey, using the former for the construction of their queen-cells. (The queen-excluder serves merely to support the cappings.) Cover up warmly whilst allowing full bottom ventilation, keep dark, and leave for about 24 hours. At the end of this time all the queen-cells the bees are inclined to make will have been well started and will be ready for transference to a cell-completing stock.

(4) After the removal of the queen-cells to a cell-completing stock the bees of the swarm-box should be shaken out in front of the stock whence they were taken.

Precautions to be observed in the use of the swarm-box.

(1) If fresh nectar has been gathered during the previous days much of it will fall on the bees when these are shaken into the swarm-box and many may be drowned in it. In this case it will be necessary to shake very lightly and to complete the dislodgment of the bees with a feather or soft brush.

(2) Unsealed queen-cells must not be exposed to low temperatures and all the foregoing operations must be performed with celerity.

(3) At no time must a cell-bearing comb be shaken. A goose wing feather is suitable for removing the bees from it.

(4) If desired the queenless bees in the swarm-box may be allowed to complete the queen-cells but in this case the box must be taken without jarring to a new position at least 1½ miles away—and the bees liberated and amply fed. On the 10th day after the insertion of the brood the cells will be ready for distribution to nuclei. Again, jarring must be studiously avoided if the swarm-box is to travel home. This course however is not generally desirable because the stock from which the bees were taken will not be doing useful work, and because the queen-cells are better completed in the supers of a powerful queen-right stock.

(5) We cannot better emphasise the two main essentials of cell-getting than by the following quotation from Jay Smith (77, p. *28*):—

"Rearing good, vigorous queens without strong colonies and plenty of feed is an impossibility. The colony that is to furnish the bees for the swarm-box must be fed at least three days before the swarm-box is filled. It will do little good to feed them just before they go into the swarm-box. I do not know why this is true, but it seems to take a few days for the bees to assimilate the food and make it over into royal jelly."

Use of funnel in filling a swarm-box.

The swarm box can be quickly furnished with bees by a skilled worker without the escape of many bees, but practically all loss is obviated by the use of a funnel as illustrated in Plate XXI. This is made of sheet metal, the top being strengthened by an exterior wooden frame. For use with British hives the upper portion has the following dimensions:—

Upper part. Length 20″; breadth, $3\frac{1}{2}$″; depth, 9″.
Lower part. Depth 6″.
Opening at base. Length, 8″; breadth, $2\frac{1}{4}$″.

The base of the funnel is fastened by retrorse extensions of the metal to a stout board of the same size as the lid of the swarm-box. The central portion of this board, corresponding to the base of the funnel, is cut away. When in use the funnel rests on the lid of the swarm-box and the two half-slides of the lid are drawn apart sufficiently to admit the bees to the box as they are shaken through the funnel. The combs are already in the swarm-box the central space under the funnel being left for the insertion of the comb

PLATE XXI

[*Photo: L.E.S.*

STOCKING A SWARM-BOX BY MEANS OF THE FUNNEL

or frame to carry the queen-cells. It is into this space that the bees fall as they are shaken from their combs. They are so disconcerted that they retreat to the combs in the swarm-box and do not attempt to escape.

Limited use of the swarm-box.

It should be clearly understood that the use of the swarm-box is to induce the bees to commence raising queen-cells and that these cells should be completed in the supers of another strong stock as described in Chapter XV. In Chapter XVII the writer describes his method of utilising one strong stock for both purposes and that without materially diminishing its activities in honey-producing.

(*C*) *Starting of Queen-cells by unconfined queenless bees. Method 22.*

Doolittle (23, p. 27) induced bees to start queen-cells in the following way:—

Remove the queen with one comb from a populous stock. Substitute a division board feeder for the removed comb. Feed with warm syrup for eight days.

Three days after the removal of the queen, remove all brood. This will contain many queen-cells in an early stage which are to be destroyed. Close up the combs containing honey and pollen but leave a space amongst them for the insertion of a frame of grafted queen-cells.

Cover the hive. Shake the bees from the brood combs in front of the hive, allowing them to run in at the entrance. Give the brood to another stock. An hour later insert the grafted queen-cells.

Great care is necessary to ensure that not a single egg or young larva is contained in the combs of stores given to the cell-raising bees.

It will be noted that the bees are not confined as in the case of the swarm-box.

Method 23.

Pritchard (67, p. *8*), queen-breeder to the A. I. Root Co. in America, recommended the following method of starting queen-cells. It is slightly different from that described by Doolittle and involves more work.

Use a stock of medium strength. Remove the queen and one of the combs, so leaving a space in the middle of the brood-nest for the reception of prepared larvae. Feed the stock with thin syrup and at the end of three days break down every queen-cell. At this time insert queen-cell cups containing transferred larvae not more than 36 hours old, taken from the best stock. As the bees must be prevented from making queen-cells on their own combs—which they are always inclined to do—the combs should be examined every other day for eight days and unwanted cells destroyed.

This method and the preceding one are good in that practically all the bees and stores are available, but larvae of less than 24 hours of age should be used. On the other hand there are the disadvantages of having to find a temporary home for the queen, if she is of value, and of leaving the stock queenless for ten days.

Method 24.

A better method used by Pritchard, and described by Jay Smith (77, p. *43*) is a modification of the swarm-box method, the bees being rendered queenless and broodless, but remaining at liberty and in their own hive. It is facilitated if a nucleus hive and a spare brood box are available.

The procedure, with slight modifications is as follows:—

Into the spare brood-box are placed:—

(1) Two combs containing honey and pollen, but no brood,—and between them.
(2) One or two frames fitted for holding cell-bars (pp. 234–235).
(3) Two division board feeders containing thick syrup, one on each side.
(4) Dummies to fill the remaining spaces.

Select a stock of medium strength, preferably without supers. Lift the brood-box from the hive floor and put it on a temporary stand. Put the prepared brood-box in its place on the hive floor. Remove the two empty (cell-carrying) frames, thus leaving a space in the centre. Into this space shake all the bees of the stock, and replace the quilts and cover (Pl. XXII).

The original brood-box will now contain the brood combs without bees. The comb of bees with the queen is added to them,—the queen being first caged—and the whole put over an excluder on a strong stock to be cared for during the ensuing 24 hours.

An hour later two frames containing the bars of prepared artificial queen-cells are inserted in the central space in the queenless stock. The bees, now in distress at the loss of their queen, and having accumulated brood food in consequence of the loss of their brood, will accept them and feed the larvae they contain. After 24 hours the bars of cells may be removed, the bees feathered off, and the bars given to a cell-completing stock (Chapter XV).

The queen and brood are then restored to their original stock.

PLATE XXII

[*Photo:* —*E.S.*

SWARM-BOX STOCKED WITH UNCONFINED QUEENLESS BEES

The trouble of caging the queen may of course be avoided if she and her comb of bees are kept in the nucleus box—closed and ventilated—during the 24 hours in which another stock is caring for the brood.

Time to introduce grafted queen-cells.

Different authorities recommend different periods of confinement of queenless and broodless bees prior to the insertion of grafted queen-cells,—e.g., Pritchard, one hour, Jay Smith, two hours, Sladen, five to six hours, and Maisonneuve, eight to ten hours. There is doubtless a time when the accumulation of secreted brood food in the nurse-bees reaches a maximum,—probably after six hours of confinement. At all events experience shows that this period is consistent with good results.

CHAPTER XIV

REARING QUEENS IN ARTIFICIAL QUEEN-CELLS

IN the preceding chapters we have described methods of queen-raising suitable for the amateur who needs queens for an apiary of limited size and who may not feel competent to undertake more scientific and complicated methods. Most of them involve uncertain acceptances, the mutilation of combs, and difficulties in cutting out and handling the matured queen-cells when ready for distribution.

For the beekeeper who needs larger numbers and a regular succession of young queens, and easy inspection and safe handling of the queen-cells, a more methodical procedure is desirable and the system commonly described as "grafting", or more correctly, "transferring" selected larvae to artificial queen cells is usually followed. We shall now describe this system in detail, and, in the next chapter alternatives to it.

It was a celebrated American beekeeper G. M. Doolittle who invented the process of making artificial queen-cups and devised a systematic procedure for their successful use. His book "Scientific Queen Rearing" was published in 1889, and as Pellett (61, p. *92*) remarks, revolutionised the industry of queen-rearing. Although certain others had anticipated some of the details of Doolittle's experiments—e.g., the transference of larvae to natural queen-cells, it was he who evolved, through a series of logical steps, the system ascribed to him, and which, in its essentials, is still used by most commercial queen-breeders.

We may here take a brief glance at Doolittle's experiments (23, p. *24*). He began by transferring, by means of a "goose-quill toothpick", selected larvae to the queen-cells of stocks preparing for swarming, having first removed from these the original larvae or eggs. He was mainly successful where larvae were substituted for larvae but less successful where they took the place of eggs,—that is to say, in dry queen-cells.

Finding that this plan did not provide him with sufficient queens, and that it was rather troublesome, he considered that if a swarming stock would adopt a larva substituted for an egg in a dry queen-cell, a de-queened stock might do the same. He collected a number of the new but unoccupied "embryo" queen-cells which are to be found in most stocks just before swarming time, stuck them on to an old and partially cut-away comb, and inserted a selected larva in each. This comb he placed in a stock immediately after removing its queen and all the brood. Next day he found that all the larvae except one had disappeared.

Reflecting on his failure he concluded that the bees were not inclined to rear young queens immediately on the removal of their queen and that an interval might be necessary after the removal of the brood for the accumulation of brood-food in the "stomachs" of the nursing bees. He therefore repeated his experiments with the following modifications:—

He removed the queen. Three days later, when the bees had started their own queen-cells, he removed all the brood, including queen-cells. An hour later he inserted the prepared queen-cells.

The next day he found them all accepted.

About this time he had made his first experiments in the making of artificial queen-cells. Taking some bees-wax made plastic by warming he moulded some round the end of a stick with his fingers, so making "a very presentable queen-cup", but he found that cups made in this way, when included in the experiments just described, were all rejected whilst natural cells "clipped off" the combs, were accepted. As the natural cells were not plentiful and were troublesome to collect he was impelled to search for a better way of making artificial ones. Remembering how his mother made candles by dipping he decided to use the same principle in making queen-cells. A forming-stick (the wooden tooth of a hay-rake) was cut down and smoothed until it exactly fitted a natural queen-cell. The end of the stick, moistened with cold water, was then dipped into melted wax to a depth of $\frac{9}{16}''$, quickly withdrawn and cooled, dipped again less deeply, and the process repeated 6 to 8 times. The result was a queen-cell with a delicately thin edge, thickening walls, and a substantial base, which was easily removed from the stick when cold.

Doolittle found that cells made in this way were not acceptable to queenless bees if given "dry" but that they were readily accepted if each was provided with a little royal jelly before the insertion of the larvae. He had now found a method of providing himself with an unlimited number of queen-cells which relieved him of the necessity of searching for natural ones, enabled him to breed at any time within season, and to handle his queen-cells with ease and safety.

A detailed description of the making and use of artificial queen-cells, based on Doolittle's method but including modifications of detail and minor improvements recommended by eminent queen-breeders and by the present writer, will now be given.

The subject divides itself into three sections, viz:—

1. The construction of the queen-cells.
2. Their attachment to suitable wooden bars.
3. The transference of selected larvae to them.

1. *Construction of artificial queen-cells.*

Queen-cups, as they are sometimes called, may be made before the opening of the season. They can be preserved indefinitely if kept from dust and a drying atmosphere. Jay Smith (77, p. *22*) recommends that they be wrapped in tissue paper and kept in foundation cartons. Before use they should be inserted into a stock for 24 hours so that the bees may clean them, shape their edges to their liking, and impart a hive odour to them. If however they are allowed to become dry and brittle, or dirty, they are likely to be unacceptable to the bees.

They must conform as nearly as possible to the shape and size of natural queen-cells, and must have smooth interiors and thin edges. They must be made of pure and preferably new bees-wax which has not been darkened by overheating.

For their construction we need:—

(*a*) Pure bees-wax, fresh from the solar wax extractor, or prepared from cappings.
(*b*) A small metal trough to contain the wax.
(*c*) Means of heating the wax so as to keep it a little above melting point.
(*d*) Forming sticks of suitable size and shape.
(*e*) A basin of cold water.
(*f*) A cloth for wiping the forming sticks.

Taking these in order:—

(*a*) It is important that only the best light-coloured wax be used. Even that from the solar wax-extractor should be obtained from the brace-combs collected during the summer and not from old combs. That prepared from cappings should be melted in hot rain-water and strained through hot thin flannel.
(*b*) Doolittle (23, p. *34*) melted his wax in a small rectangular metal trough placed on the glass chimney of an ordinary paraffin lamp which

PLATE XXIII

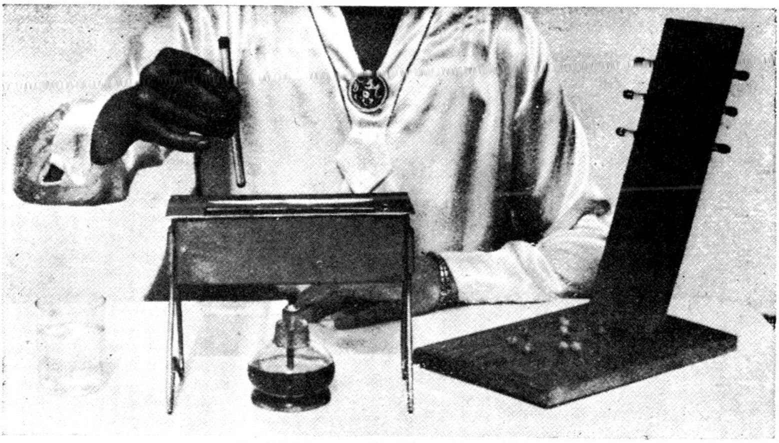

A. Making artificial queen-cells.

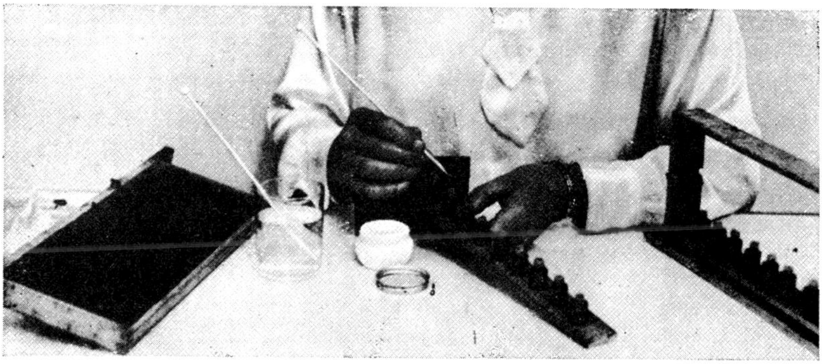

B. Grafting artificial queen-cells.

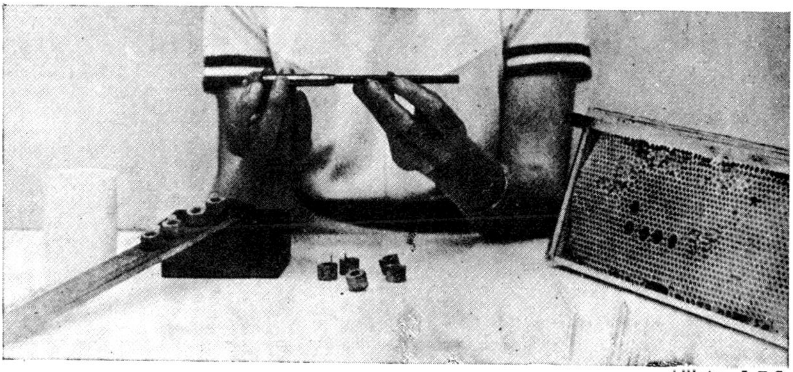

[*Photos. L.E.S.*

C. Inserting cell into cell-holder (Barbeau Method)

was so regulated as to keep the wax slightly above its melting point. He tilted the lamp sufficiently to secure that the wax was $\frac{9}{16}''$ deep at one end and of very little depth at the other end. By successive dippings of his "forming stick" along the length of the trough he obtained cell cups with thin edges and thick bases.

This rather crude and unstable arrangement, admitting of obvious improvements, might serve for the preparation of a few queen cells but for continued work something better is needed. The following is a brief description of the apparatus devised and used by the writer and illustrated in Pl. XXIII, A.

A small rectangular trough, which contains the wax, is suspended within a larger one which serves as a water bath. The latter is heated by a spirit lamp or bunsen burner placed beneath it. Both troughs are made of thin galvanised sheet iron.

The wax trough has the following dimensions:—Length, 6″; breadth, 1″; depth at one end 1½″; depth at the other end 1″.

The water trough is 8″ long, 2½″ wide, and 2½″ deep.

The wax trough, with its sloping base, is suspended by two flanges extending from the upper edges of its ends and soldered to the upper edges of the ends of the larger trough.

The superficial space between the two troughs is covered at the ends and along one side to conserve heat, and that on the remaining side is left open to admit the water. The legs are made of heavy galvanised wire and slide into cylindrical folds of metal sheets soldered to the ends. The height of the water-bath above the lamp flame can be adjusted by sliding it up and down on the legs, the tension in the latter serving to hold it in any desired position.

(*c*) When needed for use boiling water is poured into the outer trough to a sufficient depth to immerse the lower half of the wax trough. Wax is then placed in the latter so that when melted it has a depth of about $\frac{1}{2}''$ at the deep end.

The lamp flame will maintain the heat of the water and there will be no risk of overheating the wax.

(*d*) Three or four forming sticks are needed if many cells are to be formed so that the partially made wax cups may be cooling whilst others are being made.

Doolittle (23, p. *33*) used a tooth of a wooden hay-rake which he reduced towards one end to a diameter of $\frac{5}{16}''$, corresponding to that of a natural queen-cell. The end was rounded and smoothed with sand paper. A mark was made round the stick at a distance of $\frac{9}{16}''$ from the end to show the depth to which the stick should first be dipped.

It will be noted that Doolittle made his cell-cups $\frac{9}{16}''$ deep. As far as acceptance by the bees is concerned this depth is satisfactory but it is not easy for the operator to place a tiny larva on a drop of royal jelly lying at the bottom of such a deep and narrow cell so that the larva lies in its natural position (p. 241). It is quite easy to do this in a shallower cell and for this reason the writer makes cells of $\frac{3}{8}''$ depth. The bees accept these quite as readily as the deeper ones.

Pratt (66, pp. *6–10*) used a forming stick of the shape shown at 7 in Pl. XXIV. With this he made cells by compression, first filling his hollowed wooden cups with melted wax and then pressing the stick into them. The exuded wax provided extra material for the use of the bees in completing the queen-cells. He also used a simple machine (the "Grace" compressor) for forming about a dozen queen-cells at a time.

There are slight disadvantages in the use of wooden forming sticks. They absorb water and swell slightly, and wax sometimes sticks to them, especially if too hot. Box-wood is the best to use on account of its hardness, and smoothness when polished. Better forming sticks can be made from rods of bone, bakelite, or casein. They do not absorb moisture, are easily kept clean, and the wax does not so readily adhere to them.

The writer uses glass rods six inches in length and $\frac{5}{16}''$ in diameter. The ends are ground to a hemispherical shape against the side of a fine emery wheel and made smooth by heating to melting point in a bunsen flame.

To make the cell-cups:—

Place the wax in the inner trough and pour boiling water into the outer one. When the wax has melted and is $\frac{1}{2}''$ deep at one end:—

(1) Dip the forming stick into cold water, shake off any adhering drops, and insert it into the molten wax to reach the bottom near the deep end, or, if there is no graduated trough, to a depth of $\frac{3}{8}''$.

(2) Remove it at once and keeping the waxed end downwards twirl it a little with the fingers to ensure an even distribution of the wax.

(3) When set, dip it again into the wax in the next (less deep) section of the trough.

(4) Remove it quickly and again roll the stick between the fingers during cooling.

(5) Repeat these processes six times in all each time dipping to a shallower depth.

(6) Place the stick with the cell on a stand as shown in Pl. XXIII, A, to cool in the air, and proceed to make another cell with a second forming rod.

(7) Completed and cooled cell-cups are easily removed from the forming sticks by first dipping them in cold water and then applying a slight twisting and pulling movement at their bases. Care should be taken not to damage their thin edges.

(8) The cell-cups may then be stored (p. 224) or if needed for early use, attached to cell-carrying bars.

2. *Attachment of artificial cells to movable wooden bars.*

(*a*) *Cell-holders.*

The earliest artificial queen-cells were fixed into combs by means of melted wax. Later they were fixed to the under side of a wooden cell-bar after their bases had been dipped in molten wax (23, p. *37*). As these when completed were not easily handled, movable wooden holders were invented by Phillips, Pratt, Maisonneuve, and others to give them large and strong bases.

First came the simple cell-holder (Pl. XXIV, *1*), a short cylindrical piece of wooden rod $\frac{5}{8}''$ long and $\frac{5}{8}''$ in diameter. One end is slightly hollowed to receive the base of the prepared cell which is fixed with molten wax. The other end is fixed directly to the under side of a cell-bar, also by wax.

Phillips' holder (Pl. XXIV, *2*,) is similar but is more deeply hollowed at one end so as to accommodate the body of the cell. It is fixed to the cell-bar by means of a sharp metal spike projecting $\frac{1}{4}''$ from the holder.

Sladen (75, p. *8*) who used a shallow wooden dummy to which to fix his queen-cells drove the spikes (fine nails) downwards through a cell-bar and then nailed the bar to the bottom of the dummy. The spikes could not then be moved and the plain cell-holders were forced on to them.

Pratt introduced flanged holders (Pl. XXIV, *4*) the

lower portions of which, hollowed to receive the prepared cells, passed through slightly larger holes bored through the frame top or cell-bar. These are easily removed and replaced from above without disturbance of the bees.

Any of these holders can easily be turned in a lathe. Convenient dimensions for them are as follows:—

Flange—Diameter $\frac{7}{8}''$ Thickness $\frac{1}{8}''$.

Body—Length $\frac{5}{8}''$ External diameter $\frac{5}{8}''$.

Cavity—Depth $\frac{3}{8}''$ Diameter $\frac{7}{16}''$.

Maisonneuve's "demountable" holder (Pl. XXIV, *5*) is more elaborate. Made of aluminium, white-metal, or preferably of wood, it comprises a lower tube, of a size to take a Doolittle queen-cup, which opens into a larger chamber above, surmounted by a cover which is fastened to the body of the holder by a coarse screw thread or a simple bayonet fitting. It is 1″ high, and $1\frac{1}{16}''$ wide in its upper chamber. The lower tubular portion is $\frac{3}{8}''$ long and its internal diameter $\frac{1}{2}''$. Apart from its suitability to hold ordinary queen-cups, its expanded upper chamber enables it to house a queen-cell cut from the combs, and as the interior of the holder is waxed, a queen-cell may be fixed in any position, i.e., with much or little of it protruding. If only the tip of the cell protrudes the holder serves as a cell protector when the latter is introduced to a nucleus. Maisonneuve shows how this holder may be used in queen introduction and for providing food for young queens kept in nursery cages.

Pratt's flanged cups may be used with ordinary cell bars pierced to receive them or they may, as Wedmore (94, p. *56*) recommends, be inserted into holes bored in the inverted base of a shallow box $1\frac{1}{2}''$ to 2″ deep. The inverted box is placed directly on the frames of the cell-raising stock and covers such portion of the stock as is warranted by the number of artificial queen-cells used.

The flanged holders can be lifted one at a time from

PLATE XXIV

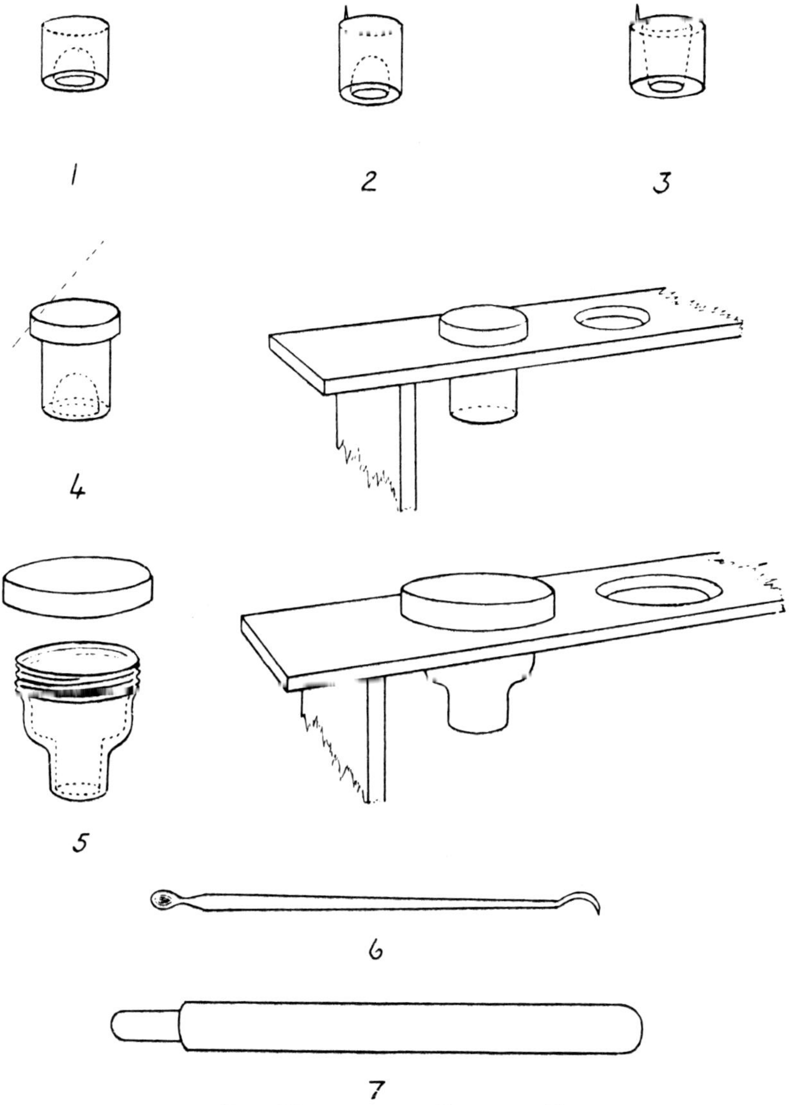

CELL HOLDERS AND GRAFTING TOOLS

Cell holders. 1. Simple; 2. Phillips; 3. Barbeau; 4. Pratt; 5. Maisonneuve.
Tools. 6. Metal grafting tool; 7. Pratt's cell-forming stick.

the top bar of a cell-carrying frame, or from the shallow box described above without disturbing the bees. When one holder is withdrawn a cork may take its place whilst it is being grafted. As this takes only a brief time the risks of chilling and drying of the larvae are reduced to a minimum. On the other hand these holders have one serious disadvantage which caused the writer, after using them for several years, to discard them. A prosperous cell-completing stock usually enlarges the exterior of a queen-cell after it is sealed by building worker comb around it, as shown in Pl. XXIX. When this happens the flanged holders cannot be withdrawn unless the queen-cells are first pared down with a knife—a troublesome and risky operation. The simplest and most satisfactory holder is that of Phillips. It is the only one used and recommended by the writer for grafting by the Doolittle method.

Alternatives to cell-holders.

The following are alternatives to wooden cell-holders:—

The lower side of the cell-bar is dipped just beneath the surface of melted wax and removed to cool. It is then dipped again and cooled, the process being repeated until the thickness of the wax layer on the bar is $\frac{1}{8}''$ or more. The bases of the prepared cells are then dipped in melted wax and attached to this layer. The cell-bases may then be painted with a camel hair brush and melted wax to fix them more securely. When mature the cells can easily be cut away with substantial wax bases.

Tinsley (86, p. 7) makes discs of honey-section wood of $\frac{3}{4}''$ diameter which he fixes by means of melted wax to the under side of the cell-bar. The cells are attached by wax to these discs which become strong bases for them when removed.

The writer has used round discs of wax, the size of

a shilling, but of $\frac{1}{4}''$ thickness. They are cut from a layer of wax of this thickness by means of a hollow metal punch first heated in boiling water. These discs make excellent cell bases, some of the wax being used by the bees for the completion of the cells.

Artificial queen-cells on combs.

Rosser (74, p. *111*) does not advocate the use of wooden cell-holders. He pushes his artificial queen cups containing larvae into the combs where they would be likely to be found in nature, e.g., at the bottoms or ends of the combs, near young larvae, and especially near the pollen cells, as bees need a great deal of pollen when queen-rearing.

Barbeau (62, p. *308*) made cell-holders of thin sheet aluminium the base of each being cut in the form of a very coarse screw. By this means he fixed his queen-cells to convenient parts of a comb and when they were mature removed them by unscrewing the holders. He then screwed each into a comb of a queenless stock prepared to receive it. Whilst it was quite easy to fix these holders to a comb, one or two turns of the screw base being sufficient, it was more difficult to unscrew them without damage after the bees had built wax about them, and Barbeau subsequently adopted the holder described on p. 261 (Pl. XXIV, *3*).

(*b*) *Cell-bars.*

Artificial queen-cells are usually attached to wooden bars which are carried in ordinary brood frames. These bars should be easily removable but fixed so that they will not fall from the frame during examination. A simple way of fixing and securing them is described on p. 235.

A British shallow frame will carry two bars spaced 2″ apart and these will accommodate about 24 cells which are as many as one can expect to be accepted in most favourable circumstances. It is true that some writers record much larger acceptations but for these very strong forces of bees are needed in the cell-starting stock. Jay Smith (77, p. *33*) for example speaks of 5 to 7 lbs. of bees in the swarm-box and an acceptance of 60 cells (100 per cent) at a time. In England however we can hardly expect to obtain so many bees from a standard brood-box and therefore anticipate more modest results. The number of prepared cells we may put into the swarm-box is of course not limited, but if too many are started the young queens will be inadequately fed by the cell-completing stock. Sladen (75, p. *11*) recommends a dozen cells as sufficient for a stock to feed at one time or from 18 to 20 for a strong stock. Doolittle (23, p. *45*) considered 12 sufficient and 24 probably too many.

The trouble usually is however not that the bees accept too many cells, but too few. Anything above a 70 per cent acceptance can be considered as good and an 80 per cent as excellent. We may therefore consider a single bar with 12 cells as a minimum and two bars with 24 cells as a maximum to be given to a stock at one time.

It is a common practice to use the top bar of a frame as one of the cell-bars especially when Pratt's holders are used. The cells can be inserted, examined, or removed with negligible disturbance of the bees. If a second bar be used, in a lower position in the same frame, a better acceptance may be expected in it than in the top bar. In nature the bees build most of their queen-cells in the lower regions of the combs, and seldom near the top bars. The writer prefers not to use the top bar of the frame but instead to place a movable bar $\frac{3}{8}''$ below it so that the bees can move freely above as well as below the cells (Pl. XXV). Sladen's cell-bar attached to a shallow dummy, whilst ensuring the best position for the queen-cells does not permit the bees to circulate over the cells, and as Herrod-Hempsall (33, I, p. *644*) observes, it divides the cluster of bees.

It could be improved by a bee-way between the bar and the dummy.

In all cases it is preferable to mount the cell-bars in a shallow rather than a deep frame. They can then be inserted into either a shallow or a deep super of a cell-completing stock. Although an ordinary frame can easily be adapted to carry cell-bars it is desirable to make a special shallow frame (Pl. XXV) having the following dimensions:—

Length of frame, the same as for an ordinary standard frame.
Depth of frame, $5\frac{5}{8}''$.
Width of all parts, including cell-bars, $1\frac{1}{4}''$.
Thickness. Top bar, $\frac{3}{8}''$; side bars, $\frac{1}{2}''$; bottom bar, $\frac{1}{4}''$.
Length of cell-bars, $13\frac{3}{8}''$.
Thickness of cell-bars, $\frac{1}{4}''$.
Distance between top bar and upper cell-bar, $\frac{3}{8}''$.
Distance between two cell-bars, $2''$.

The cell-bars slide easily into grooves cut $\frac{1}{4}''$ deep in the side bars. They are secured by small metal buttons on both sides of the frame and cannot then fall out when the latter is turned or reversed for inspection of the queen-cells.

If a brood frame be used, the space above—not needed for the accommodation of the bars and cells—may be filled with foundation or comb.

A cell-bar for a British standard frame will accommodate 12 queen-cells spaced about 1″ apart. It is better to limit the number to 10 as the bees are then less likely to join them together with brace-comb. In times of heavy income the bees may almost bury the queen-cells in comb but they will always leave the ends clear so that the young queens may emerge. In such cases the almost buried queen-cells when ready for removal should be carefully cut out with a knife and trimmed with small scissors.

3. *Transference of larvae to artificial queen-cells.*

(*a*) *Grafting equipment.*

We now have to consider the process of transferring the selected larvae to the prepared artificial cells. The operation, although a delicate one, is extremely interesting and becomes easy after a little practice. To ensure success we must observe certain conditions. The tiny larvae with which we have to deal are easily injured, chilled, overheated, dried, or starved, and it is essential that they be not kept out of a hive longer than is absolutely necessary. Environment and equipment must therefore be prepared beforehand so that we may work expeditiously.

We shall need the following:—

(1) A room (e.g., a bee-house, the honey-house, or the kitchen), with suitable conditions of light, temperature, and moisture.
(2) Cell-bars with the queen-cups attached to them (p. 233).
(3) The frame to take the cell-bars (p. 235).
(4) The comb containing the selected larvae, kept in a warmed box.
(5) A sharp knife or razor.
(6) A grafting tool.
(7) Some royal jelly.
(8) A tiny spoon or additional grafting tool for lifting royal jelly.
(9) A vessel of hot water.
(10) The swarm-box with confined bees.
(11) Feeders and coverings for the swarm-box.
(12) A warmed nucleus box (p. 239).

The room in which grafting is to be done should have a temperature of about 70° F., and the air should be moist. These conditions can be attained by placing boiling water over a small stove of suitable size and keeping in the room for some time before operating.

The grafting tool must be suitable for lifting a tiny larva from its original cell and depositing it on a little royal jelly in the prepared artificial cell, without injury, without abstracting heat from it, and without materially altering its shape and position relative to the cell. We may use any of the following:—

(*a*) Doolittle's quill, the shaft of a feather cut as though for a quill pen, the sharpened portion being 1″ long, and the tip $\frac{1}{16}$″ broad and turned slightly upwards.

(*b*) Maisonneuve's "thin wooden blade" or Herrod-Hempsall's match stick, shaped as above.

(*c*) The combined metal spoon and curve-ended tool as sold by manufacturers (Pl. XXIV, *6*).

(*d*) The tool used by the writer (Pl. XXIII, B) is fashioned as illustrated from a casein knitting needle by means of a round file and fine emery cloth.

The metal tool has a smooth and suitably curved end. As a good conductor of heat it quickly abstracts heat from a warmer body and so may chill a larva with which it comes in contact. The quill, match stick, and casein tool do not conduct heat well and therefore are not likely to chill the larvae. The quill and match stick may have edges sufficiently rough or sharp to wound larvae. The casein tool is made perfectly smooth by the emery cloth.

The tool should be warmed in tepid water and dried immediately before use. Herrod-Hempsall (33, I, *641*) recommends that this be done in the operator's mouth.

(*b*) *Provision of royal jelly.*

Royal jelly can be obtained at any time by making a stock queenless for three or four days. From the resulting queen-cells the larvae are removed and the cells containing royal jelly taken to the grafting room for immediate use. Small portions of the jelly are taken from them as required for the queen-cups.

Alternatively the jelly can be scooped from all the cells and stored in a small air-tight vessel,—e.g., a shallow wide-mouthed jar with a screw top or tight-fitting cork. It will remain good for months if kept from air, light and drying. Jay-Smith (77, p. *23*), relating how he collects royal jelly from unwanted queen-cells throughout the season and stores it in a small screw-capped porcelain jar which he keeps in his pocket, states that the idea of preserving the jelly in this way originated with J. W. George.

As royal jelly is more dense towards the bottom of a queen-cell than at its surface it is customary to stir it to a uniform consistency before use although it is probable that the thin surface layer is more suitable for a very young larva. It should be of the density of "a thin cream" (33, I, p. *638*) and if necessary may be brought to this condition by dilution with one or more drops of distilled water or clean and previously boiled rain water. If it is too dense or too thin the bees are likely to consume it and remove the larvae. Herrod-Hempsall advises that in cool weather stored royal jelly be slightly warmed by standing the vessel containing it in warm water before use, as cold jelly may kill the tiny larvae.

These are fine points but worthy of observance if we desire the maximum of success.

The amount of royal jelly needed for each artificial queen-cup is very small and is variously described, e.g..

Doolittle (23, p. *36*)—"the size of a BB shot, or a drop $\frac{1}{8}''$ in diameter".

Cowan (20, p. *128*)—"as much as will go on the point of a pen knife".

Herrod-Hempsall (33, I, p. *639*)—"about half the size of a mustard seed".
Sladen (75, p. *10*)—"the size of a no. 2 shot".
Jay Smith (77, p. *37*)—"about twice the size of a pin-head".

The royal jelly spoon may be that supplied by the manufacturers (Pl. XXIV, *6*), or may easily be made from a strip of tin. It should be kept dry and slightly warm whilst in use and the jelly pushed from it into a cell-cup by the grafting tool. The latter may then be used to spread the jelly evenly over a part of the base of the cell. The writer does not find it necessary to use a spoon. A second casein grafting tool (Pl. XXIII, B) is more convenient both for lifting and depositing the jelly.

When all the necessary equipment is ready and within convenient reach, and not before, the comb of selected larvae, without bees, is brought from the hive into the operating room. To avoid risk of chilling it should be carried in a closed nucleus box previously warmed in the manner described below. This comb should contain eggs, and larvae of the first and possibly the second day and if the stock has been liberally fed the larvae will be well provided with jelly for their age. The comb is laid flat on the table and a patch containing larvae from 12 to 24 hours old (p. 66) is selected. With a razor, dipped in hot water and dried, the cells of the selected patch are sliced off to within $\frac{1}{8}''$ of the larvae to make these easily accessible to the grafting tool. Alternatively and to avoid mutilation of the comb, a line of cells containing suitable larvae is chosen and the cell walls pushed apart in opposite directions by the blade of a table-knife (33, I, p. *641*) (Pl. XXIII, B).

(*c*) *The warmed nucleus box.*

Precautions against the chilling of larvae and the drying of royal jelly during grafting operations, and the chilling of matured queen-cells when these are being distributed to nuclei are seldom or never referred to by writers on the subject of queen-rearing. Jay Smith however (77, p. *37*) keeps a wet towel over his grafted larvae to keep them from drying pending their transfer to the cell-raising stock. The writer has eliminated all risk of chilling or drying by the following simple expedient:—

A nucleus box of a size to accommodate four or five deep frames is provided with a close fitting cover. An ordinary rubber hot-water bottle filled with hot water is laid on the bottom and this is covered by a towel which is first dipped in very hot water and then wrung out. The lid is then closed. The box quickly fills with vapour and a temperature of 70°F. to 80°F. can be maintained in it for a long time.

The box accommodates shallow frames which hang clear of the hot water arrangements. If full frames are used for queen-rearing the box must of course be deeper.

The box is used

(1) When taking a comb of larvae from the breeding hive to the grafting room.
(2) To accommodate the frame carrying the cell-bars, and to receive each cell-bar immediately the cells are grafted.
(3) To carry started queen-cells to a cell-completing stock.
(4) To carry matured queen-cells to nucleus hives (Pl. XXXII). When used for this purpose it is provided with a board the length of the top bar of a frame and about 4″ wide and $\frac{3}{8}$″ thick. This serves as a wide top bar, and saw-cuts, five on each side, accommodate the vertical wire extensions of the queen-cell protectors (pp. 294-295).

The lid is closed immediately after the insertion or removal of a frame. In the illustration (Pl. XXXII), it is raised only to show the method of suspension of protected queen-cells.

Process of "Grafting".

The transference of a very young larva together with a little of its surrounding food to a new cell is perhaps the most delicate operation in practical beekeeping. It demands good sight and a steady hand, but after a little practice becomes surprisingly easy. A beginner does well to get preliminary practice by transferring young larvae from one worker cell to another and should become accustomed to recognising larvae of different ages. In this the use of a low power magnifying glass is at first helpful.

Having everything ready we proceed as follows:—

Sit at a table in such a position that the best light falls on the prepared cells and larvae. Place the cell-bars, with cells affixed, upside down on the table and with a spoon or grafting tool place in each cell the requisite small amount of royal jelly (p. 238). Spread it over a portion of the cell base with the grafting tool. Twenty or more cells can be treated in this way in about 5 minutes.

Without loss of time place the tip of the grafting tool under a selected larva so as to lift with it a small portion of its surrounding jelly. Try to lift the larva so that only its curved back rests on the tool,—this because it can be the more easily slipped off. Place the tip of the tool lightly on the royal jelly in the artificial queen-cell and slightly tipping it sideways until the ends of the larva touch the royal jelly, float it off by gently drawing the tool sideways away from it. The larva should then float on the royal jelly, its curved shape unaltered, and in the same position relative to the walls of the cell which it originally occupied. This operation should not take more than half a minute. Stock the other cells in the same way and put each cell-bar when complete into the bar-holding frame which is to be kept in the closed warm box.

The time taken for these operations is halved if an

assistant inserts the royal jelly into each cell before the operator places the larva in it.

If one person performs both operations the royal jelly may show signs of drying before they are completed. If this is observed the tool with which the jelly is transferred should be used wet. Its end should be dipped in warm water and the adhering drop, if any, shaken off. The film of water left on the tool will be sufficient to dilute the jelly.

Inserting the grafted cells into the swarm-box.

When grafting is complete it is necessary to insert the cell-carrying frame into the prepared swarm-box (p. 212). The space left for its reception will be filled with clustered bees which will escape in great numbers if the box be opened with the aid of a smoker. If however the box be first raised a few inches from the floor and lowered to it with a sharp jerk the clustered bees will fall to the hive floor. The cover may then be quickly removed and the cell-carrying frame inserted with little loss of bees. A simpler and more effective method however is to use the carbolic cloth as described on p. 272.

After the insertion of the frame provide feeders and coverings and keep the box in a suitable place for 24 hours as described on p. 212.

In all these operations our chief concern must be that the larvae are not exposed to low temperatures or the royal jelly allowed to dry. At every stage therefore the moist warm box should be in use and kept closed when possible.

Double grafting.

Reference has already been made to the interruption of the natural feeding of the larvae which is involved in grafting (p. 71). This may be harmful in some cases and may be the cause of rejections. Some breeders therefore practise what is known as double grafting.

PLATE XXV

[*Photos: L.E.S.*

GRAFTING METHOD IN THREE STAGES

The first grafts are given to the swarm-box for 24 hours and are then transferred to the cell-completing stock (Chapter XV). After 24 hours in this stock the cells that are accepted will be well provided with royal jelly. The larvae are then removed from them and younger larvae from the breeding stock carefully put in their places. All will be accepted, and there will be no foodless interval.

CHAPTER XV

THE CELL-COMPLETING STOCK

So far we have shown how selected eggs and young larvae are obtained and how specially prepared queenless bees may be induced to commence queen-cells over them and to nourish the young queen-larvae during the first 24 hours of their existence.

As shown on p. 66 these larvae must be provided by the bees with abundant royal jelly for an additional four days. We might of course allow the cell-starting bees to complete the queen-cells, but this would involve absence from their brood and queen during a time when they might be contributing to the honey harvest. Moreover their conditions of income, temperature, and capacity to provide royal jelly in quantity, are not as good as in a powerful stock working normally with its queen, brood, and natural stores.

It is convenient here to make reference to Doolittle's discovery (23, p. *40*) that a stock engaged in superseding its queen will accept and complete grafted artificial queen-cells provided that its own queen-cells are first destroyed. Doolittle relates that he succeeded in obtaining several sets of finished queen-cells from such a stock, and that these were excellent in every respect.

He had observed that when young brood is raised above an excluder the bees usually build a small number of queen-cells on it, and he concluded that they did this under the supersedure impulse although the queen below the excluder might be quite good. He therefore reasoned that the supersedure impulse might be induced in the supers of any powerful stock by raising some young brood above the excluder, and that such a stock would accept and complete queen-cells already started in a cell-raising stock. His experiments proved successful and he found that not only could he raise queen-cells in the supers of a stock without interfering with the honey harvest, but that he could secure an acceptance of a fresh set of prepared cells when the first set were sealed but not yet removed; and that some were accepted, although in diminishing numbers, if a fresh set were provided every two days; that is to say he got queen-cells in all stages of development in one super at the same time. As the first batch were better provided with royal jelly than later ones Doolittle decided that it was better to await the removal of the first set on the 10th day before presenting the bees with a second task.

We may here state that experience shows that a second batch of cells is seldom received as well as the first unless an interval of several days elapses after the removal of the latter.

Essential conditions of a good cell-completing stock are:—

(1) The stock must be very strong and crowding all its combs.
(2) At least two supers must be occupied. If there are more, so much the better.
(3) The stock should have been fed for at least three days prior to use unless honey is coming in plentifully.
(4) On the day before use the topmost super should be furnished with
    (*a*) Combs with honey.
    (*b*) A comb containing pollen.
    (*c*) Two combs containing unsealed brood (not eggs or young larvae), one to be on each side of the central space reserved for the reception of the started queen-cells. These combs will attract nurse-bees from the brood-nest below (23, p. *43*).

PLATE XXVI

[Photo: L.E.S.

EXAMINING SWARM-BOX FOR ACCEPTANCES BEFORE TRANSFERRING TO CELL-COMPLETING STOCK

We may now consider how we shall secure the desired conditions.

(*1*) *Using a stock supered in the ordinary way.*

In this case we have to wait until the season is fairly well advanced before undertaking queen-rearing, and we may have to negotiate the difficulties associated with swarming.

The most suitable stock for our purpose is one whose brood chamber has been supplemented by the addition of a rack of shallow worker combs. The enlarged brood-nest tends to delay swarming and if the hive is kept in the shade, amply ventilated underneath, and supered with drawn combs, swarming may possibly be avoided.

When the bees fully occupy two supers above an excluder the stock is suitable for cell-completing. It should be fed, if necessary, with a little honey and water, for three days before queen-cells are given to it.

The shallow-comb rack forming the upper part of the brood-nest should contain honey, pollen, and brood, and may easily be prepared for the reception of the started queen-cells. We proceed as follows:—

Remove the rack to one side and look over the combs for the queen. If she is found place her in the lower brood-box. If not, gently shake the bees from their combs into the lower brood-box. Remove all combs containing eggs or larvae less than 3 days old. These are to be given to another stock. Select two of the shallow combs, containing older unsealed larvae, and place them in the middle of the rack with a one-comb space between them. Retain the combs with stores and sealed brood and fill the remaining spaces with dummies. The prepared rack is now ready to be placed over a second excluder at the top of the hive.

The whole hive then stands re-constituted as follows, starting from the floor-board:—

Brood-box, excluder, two honey supers, second excluder, prepared rack, quilt, feeders, more quilts.

We must remember that if we leave eggs or young larvae in the prepared rack the bees will tend to raise queen-cells over them and will correspondingly neglect some of the prepared cells presented to them. If a single queen-cell is allowed to be built on the original combs acceptance will be disappointing. We may here observe that one modern writer—perhaps through inadvertence—recommends the inclusion of "young" brood in the cell-completing rack.

It will be noted that the effect of withdrawing the combs of eggs and young larvae and substituting dummies for them is to congest the bees so that the prepared rack becomes crowded with them,—an essential condition of success.

Should the requisite two combs of older unsealed brood not be found in the rack to be prepared they should be obtained from another stock.

The stock is left in its reorganised state for 4 to 6 hours so that the bees have time to re-distribute themselves and to experience the desire to supersede their queen. The started queen-cells are then removed from the swarm-box or other cell-starting hive and inserted in the space provided for them. Feeding should be continued for at least five days unless nectar is coming in freely, and the hive should be warmly covered with quilts.

If larvae of the first day are used for the swarm-box they will be two days old when placed in the cell-completing stock. Three days later they will be sealed, and will be ready to be removed about five days later, i.e., ten days from the hatching of the eggs. In emergency they may remain until the 11th day but only if there has been no miscalculation as to their age. If left longer one may emerge and destroy the others.

(2) *Using a built up stock early in the season.*

If we desire a cell-completing stock for use at the time of an early nectar flow, say the last week in May, we may proceed in the following way:—

Choose a strong stock with a vigorous queen and stimulate it by slow feeding throughout the spring months. As soon as the brood-box is full withdraw from it two or three combs of sealed brood, without bees. Substitute for them empty worker combs or foundation.

Place the withdrawn brood-combs in a second brood-box and use this as a super over an excluder. The empty space may be temporarily occupied by empty combs or foundation.

About five days later examine the raised brood-combs to make sure there are no queen-cells on them, and then raise one, two, or three more combs containing sealed brood and without bees, from the brood nest to the super. Transfer a corresponding number of the broodless combs from the super to the brood-box.

Repeat this process every five days, raising sealed brood, and preventing queen-cells from being formed over undetected eggs. The amount of brood which may be raised on each occasion will depend on income and weather.

A stock so treated will increase rapidly, will be unlikely to prepare to swarm, and will store well if a nectar-flow occurs. To hasten development sealed brood may be added from other stocks.

As soon as there are two full supers of raised combs, with most of the brood emerged, the stock will be in excellent condition to serve as a cell-completing stock. The preparation of the upper super for the reception of started cells is similar to that described on p. 247, viz:—

A central space is flanked on each side by a comb of maturing unsealed brood and an excluder is inserted between the supers. The started cells are given after a few hours and feeding is continued until the queen-cells are removed.

(3) *Uniting two stocks.*

A strong stock can be more quickly obtained by uniting two stocks by the newspaper method. An excluder is placed over the stock whose queen is to be retained, and an undamaged sheet of newspaper placed over the excluder so as completely to cover it. The second stock, its queen removed, is then placed on the newspaper. All this should be done in the late evening. The united stock is then increased as described in the previous method until there are two full brood-box supers over the original brood-nest. Special care will be needed in this case to remove queen-cells which may be started in the supers during the first week after uniting.

(4) *Building up a strong stock for cell-completing after the honey-harvest.*

It is not easy to build up a powerful cell-completing stock after late July, when brood-rearing and sources of income are diminishing. At this time however the hives from which the honey has been taken are overflowing with masses of unwanted bees and we may with advantage use some of these for our purpose.

The present writer runs a number of small out-apiaries and brings surplus bees from them to a selected stock at home. The beekeeper who keeps all his hives at home might take his selected stock to a distance of about 2 miles, together with the bees to reinforce it and after an interval of about 10 days flying weather, bring the united bees home again. The writer's procedure is as follows:—

A strong stock at home is selected, or is brought from an out-apiary. To it are to be added two or more boxes of bees obtained from out-apiary stocks. These are obtained as follows:—

Over any stock recently deprived of its honey place an excluder and on this a box of recently extracted

PLATE XXVII

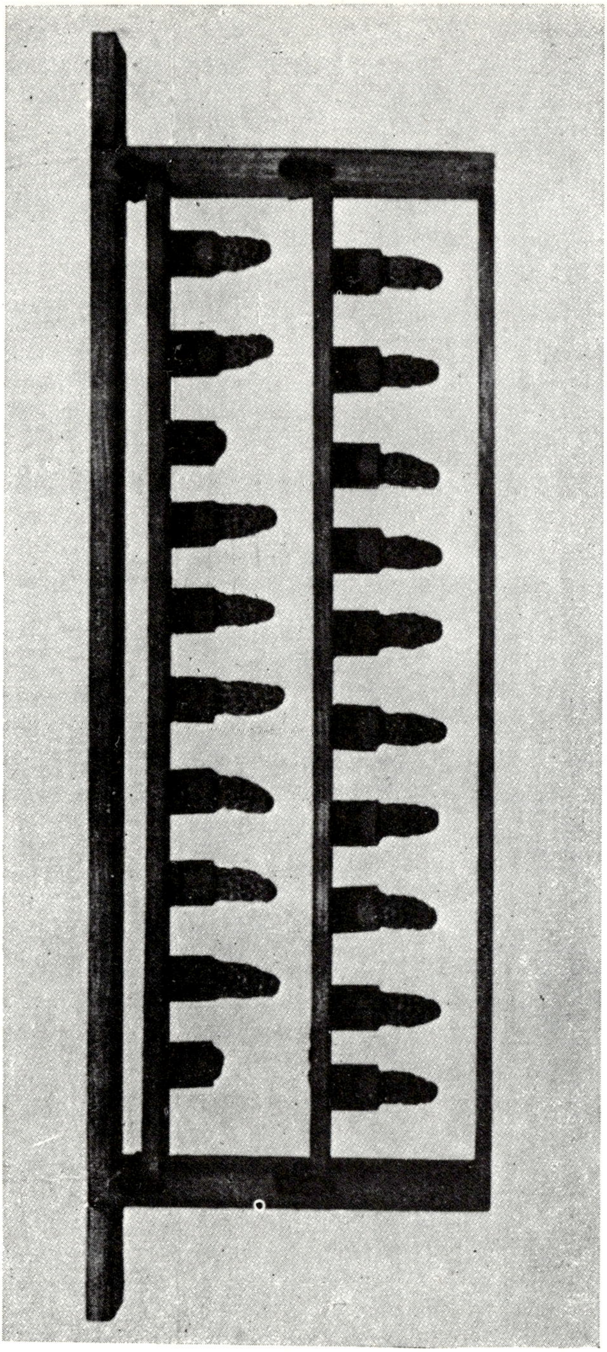

[*Photo: L.E.S.*

COMPLETED QUEEN-CELLS. DOOLITTLE METHOD

combs and two or three combs containing some maturing unsealed brood, taken without bees from the brood nest. Do the same with a second similar stock After an hour or two these boxes will be filled with bees (without drones) and a good proportion of nurse bees will have risen to the brood. Each box is then lifted on to a screen-board made of a wooden rim covered with a sheet of perforated zinc. A similar screen is placed on the top of the box, and both screens are secured by carpet staples.

The boxes are then taken home, and at evening are united to the selected stock at the same time, one above the other, by the newspaper method. In this case two sheets of newspaper will be needed. Two days later the frames containing the added brood are assembled in one of the boxes. The bees are shaken from the remaining added combs and these with the second box are removed. The box containing the brood will then be thickly crowded with bees and when placed over an excluder will be suitable for cell-completing. In this way a stock may be increased to any degree of strength and will be ready for cell-reception as soon as the raised brood is too old for the raising of queen-cells.

Using a stock preparing to swarm.

A stock making preparations for swarming is excellent for cell-starting or cell-completing, or both. It should be deprived of its queen, queen-cells, and all its young brood. These may be removed to form a nucleus. We must not lose sight however of the fact that the swarming impulse will persist in both stock and nucleus and we must take appropriate preventive measures. The nucleus will lose its flying bees to the stock and this will generally cause it to break down its queen-cells within 5 to 7 days. If this does not happen by the 5th day the nucleus should be caused to lose its flying bees a second time.

The stock itself will complete successive batches of queen-cells and will not swarm as long as the queen-cells are removed before any queen emerges. When no longer needed for cell-completing an interval of 10 days without queen-cells or young brood will cause the bees to lose the desire to swarm and they may then be re-queened, or re-united to the nucleus.

Raising queens for a continued supply.

The raising of queens on a large scale does not involve any new principle but demands the employment of more cell-starting and cell-completing stocks, to be used in carefully planned succession, and the provision of sufficient nuclei for the mating of the young queens. It is easy to produce many queen-cells but is useless to have more than can be received into nuclei. Moreover we should know beforehand how we are going to dispose of the young queens when fertile. During the swarming season or a time of nectar flow a nucleus with a fertile queen will develop rapidly, and if not relieved, will swarm. It is at these times that the ordinary beekeeper does not wish to disturb his honey-storing stocks for the purpose of re-queening.

The professional queen-breeder markets his young queens when fertile and tested (p. 312) and so makes room in the nuclei for their successors.

The prudent queen-rearer should therefore reason backwards, so to speak, and ask himself the following questions:—

(1) How many queens do I need?

Allowing for failures of acceptance, loss of virgin queens in mating flights, and possible accidents, not more than 50 per cent of prepared queen-cells should be considered likely to result in fertile queens.

(2) Can I provide sufficient nuclei to receive all the matured queen-cells?

In favourable circumstances a nucleus is in use for at least three weeks before the first brood of a young queen is sealed, and the queen ready to be removed. If she is to be tested she should remain for an additional fortnight. These periods may be shorter or longer according to the weather. The practised beekeeper can recognise worker brood before it is sealed by its amount, regularity, and continuity, and can therefore save time by removing the queen earlier.

For the purpose of approximate computation therefore we should allow an average period of a month as necessary for the rearing of a queen after the insertion of the queen-cell.

(3) Can I dispose of the fertilised queens so as to make room for their successors, or will fresh nuclei be necessary?

The economical management of nucleus hives is described in Chapter XVIII.

Causes of failure in artificial queen-rearing.

It may be convenient here to consider the usual causes of failure in artificial queen-rearing. Maisonneuve (62, p. *442*) summarizes them as follows:—

(1) Queen-rearing too early or too late in the season when nurse-bees and food may be scarce. In these cases the bees may destroy the queen-cells given to them.
(2) Some seasons are more unfavourable than others for the raising and mating of queens.
(3) Grafting out of doors or in an insufficiently warmed room.
(4) Queen-cells given to bees prematurely, or without precautions.
(5) Natural worker cells intended to become queen-cells not enlarged (p. 263).

(6) Larvae less than 12 hours or more than 36 hours old selected for grafting.

(7) Larvae chilled during transference.

(8) Insufficient wax available for the construction of queen-cells.

(9) Cells fall from the cell-holders.

(10) Cells cut from comb containing honey on the reverse side. (The bees consume this honey and may nibble through the base of a queen-cell after doing so.)

(11) Cells used for grafting a second time not previously cleaned out.

(12) The presence of young brood or an unnoticed queen-cell in either the cell-starting or cell-completing part of a stock when grafted cells are given. (From the 8th day after the last egg was laid the bees cannot raise queen-cells of their own.)

(13) Irritation of bees (by thunder-storm, bad manipulation, or hot smoke) when queen-cells are given.

(14) Giving queen-cells to be completed above an excluder when there is little or no income.

(15) Giving queen-cells before bees have had time to become hopeless after the loss of their queen.

(16) Some stocks are more reluctant than others to rear queens.

(17) Queen-cells are carelessly inserted into nuclei in such a way that their ends touch combs to which the bees attach them, so preventing the queens from emerging.

To these we would add:—

(18) The drying of royal jelly used in grafting.

(19) Injury to a larva during transference.

(20) Drowning of the larva in royal jelly.

(21) Giving grafted cells to hungry or thirsty bees.

(22) The use of defective excluders.

(23) Clumsy manipulations causing damage to queen-cells.

(24) Exposure of queen-cells at any stage to low temperatures.

Causes of failures in the queen-mating apiary are considered in Chapter XX.

CHAPTER XVI

THE "BARBEAU" METHOD AND ITS MODIFICATIONS

As the reader of the preceding chapter will have realised the "Grafting" method of providing queen-cells is not without disadvantages. Time and equipment are needed for making the cells, and the transference of a fragile one-day old larva to a fresh home demands practice and skill on the part of the operator. The disturbance of the larva is unnatural and there are risks that it may die of wounding, chilling, drying, drowning, or interrupted nutrition. The method is therefore suitable only for the experienced and skilled breeder and not for the novice.

Hopkins (35, p. *195*) who established and managed a State queen-rearing apiary in New Zealand ascribed heavy losses of queens in the post from America to the fact that they had been bred by the grafting method which in the first half of the "90's" had become fashionable. He contended that the losses were due to weakness due to the cell-grafting system of breeding.

Muzzati (55, p. *61*) cites Jay Smith as writing that "grafting" of larvae has many advantages but that if the quality of the queens produced be alone considered it is anything but a good method even if we admit that it does not harm the larvae, especially the young ones.

Muzzati, convinced of the desirability of raising queens from eggs, transferred them with great care to artificial queen-cells, but no larvae hatched from them.

He refers to Goosev, who in 1860 was probably the first to transfer the bases of worker cells with eggs into artificial queen-cells.

There are many advocates of breeding from eggs as

PLATE XXVIII

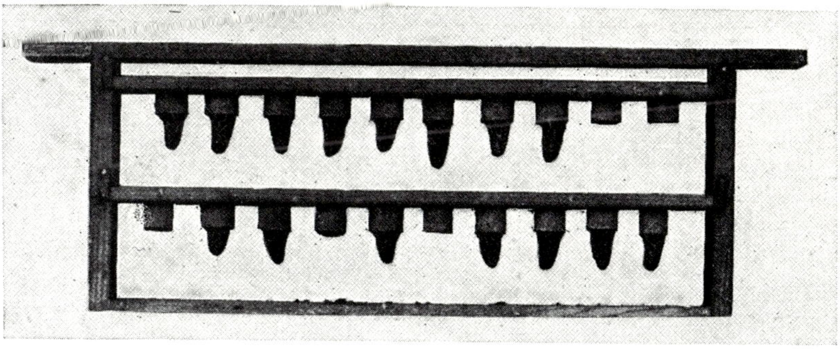

A

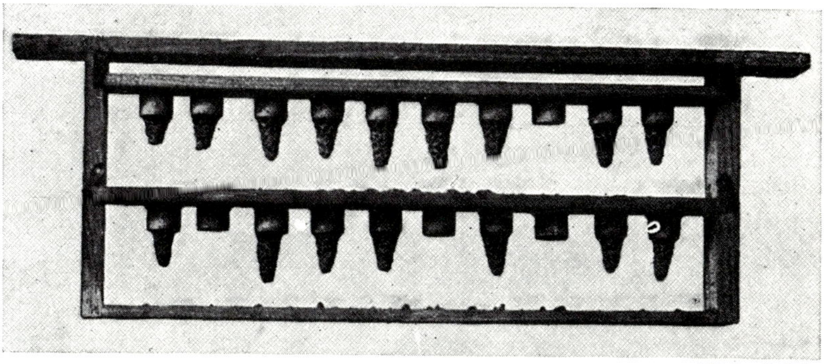

[*Photos: L.E.S.*

B

BARBEAU METHOD

A. Cells punched from old comb.
B. Cells punched from new comb.

distinct from young larvae. The advantages are obvious, —appropriate nutrition throughout the larval stage of the young queens is ensured, and neither the bees nor the beekeeper can select larvae which are too advanced to become good queens. When therefore queenless and broodless bees are caused to raise queens on a comb, eggs should be provided instead of larvae, but if the queens are to be bred in artificial cells, as can be done by the method described below, care should be taken that the eggs are of the third to fourth day, i.e., when they are about to hatch (p. 33). If the eggs are of different ages the oldest will be accepted whilst the younger ones will often be removed.

Any method by which the disturbance of the egg or larva in its original cell is avoided, and which at the same time permits of the resulting queen-cell being built in a movable cell-holder is to be preferred to the method of grafting.

Such a method was devised by E. Barbeau, a French-Canadian, and published by C. Vaillancourt in the "Journal d'Agriculture" of Quebec in March 1918 (62, p. *394*). The original article is given in full by Maisonneuve, pp. 395–399.

Barbeau used the following simple apparatus:— (Pl. XXIII, C).

(1) A cylindrical punch, e.g., a short length of steel tubing, $\frac{3}{8}''$ in diameter, sharpened to a knife edge at one end.

(2) A longer cylindrical wooden rod fitting easily into the punch.

(3) A cell-holder made of thin sheet aluminium, conical in shape, the smaller end having an aperture equal to that of the punch. The larger end was so cut as to serve as a coarse screw.

Briefly, the purpose of the punch was to cut from a selected comb a single worker-cell containing a one-day-old larva. The wide end of the cell-holder was then held against the blunt end of the punch and the excised cell gently pushed along the tube of the punch and into the cell-holder until it protruded from the latter to such an extent that only $\frac{1}{8}''$ of its length remained within the holder. The walls of the protruding cell were then broken slightly apart so as to expose the undisturbed larva. The holder, with its cell, was then screwed into the face of a comb of a stock prepared to raise queen-cells.

The inconvenience of severing these screwed-in holders from the combs was realised by Maisonneuve who inserted the cells into his "de-mountable" cell-holders (Pl. XXIV, *5*), and subsequently Barbeau himself inserted them into wooden holders of the Pratt type (Pl. XXIV, *4*) a tapering passage being bored right through them (62, p. *410*). Herrod-Hempsall (33, p. *651*) used a simpler holder of the Phillips pattern (Pl. XXIV, *3*) $\frac{7}{8}''$ long, through which a $\frac{3}{8}''$ passage was bored. This passage was tapered slightly by widening it with a file towards the upper end without enlarging the opening at the lower end. The holder was attached to a cell bar by means of a sharp spike projecting from its upper rim.

The writer uses holders $\frac{5}{8}''$ deep and $\frac{3}{4}''$ wide. The tapering passages are easily bored in a lathe, the top slide of the lathe being set for taper boring.

A suitable punch may be made from a 3″ length of plated steel tubing such as is used for towel rails or curtain rods. The internal diameter should be $\frac{3}{8}''$ and one end ground to a sharp cutting edge by means of an emery wheel. The smooth wooden rod fitting easily into the tube may be 6″ long. An ordinary lead pencil is quite suitable (Pl. XXIII, C).

Proceed as follows:—

Place the comb of selected eggs and young larvae on a flat sloping surface facing the best light. Decide on the portion of the comb which contains larvae of about 24 hours of age and select neighbouring larvae in succession so as not to mutilate the comb unnecessarily.

Warm the punch in moderately hot water, place the cutting end over the selected cell, looking down the tube to make sure that the cell is central. With gentle pressure and a screwing motion press the punch right through the comb and withdraw it. The cells on the opposite comb face will be a little crushed but that does not matter. If the punch has been carefully placed the selected cell will be uninjured and its contained larva undisturbed, and the whole will remain in the tube of the punch.

Place the cell-holder with its wider opening at the blunt end of the punch holding them together with the left hand so that they form a continuous tube (Pl. XXIII, C).

With the right hand insert the wooden rod into the punch at its cutting end and gently push the cell along the tube and through the holder until it projects from the narrow end of the latter. About one-third of its length, the part containing the larva, should remain within the holder. The protruding portion may be slightly shortened with small scissors and the shape of the mouth of the cell restored with a grafting tool.

Proceed in the same way to stock the requisite number of holders. Before fixing them to the cell-bars, it is as well, in order to ensure that none of the cells fall out when suspended in the hive, to put a little melted wax into the upper opening of the holder by means of a camel-hair brush. Similarly, when the holders are fixed to the cell-bar, if there is any doubt about the spikes holding, a little melted wax painted round the junction of holder and bar will make all secure.

Before placing the bars in the frame (here we are following the directions of Barbeau) place them on the table, cells upwards, and carefully break down the projecting cell walls by pressing them apart and downwards so as to expose the contained larvae. Barbeau considered that if this were not done some of the larvae continued to be fed as workers. The broken cell walls are removed by the bees when they begin to construct the queen-cells. The writer however has not found it necessary to break the cell walls. As previously stated, if they project from the holders much more than $\frac{1}{8}''$ they are shortened with small scissors and their cut edges restored to their normal shape with the grafting tool. In no case within the writer's experience has a larva been reared as a worker.

Fix the cell-bars in the holding frame and place the latter in the cell-starting stock (Chapter XIV).

Instead of being warmed the punch may be slightly lubricated by wiping it both inside and outside with a piece of vaselined cloth. When the punch is pressed into the comb it cuts through the six cells surrounding the selected cell. It is as well to inspect the chosen cell when mounted to see that no additional larvae have come through the tube.

The whole operation takes much less time than it does to write about it, and as no artificial cells or royal jelly are needed the amount of labour and time involved are far less than in the case of the grafting method. The larvae are not likely to be chilled nor is their nutrition interrupted, and consequently the proportion of acceptances is usually high.

It is not advisable to use cells containing eggs for this method unless these are on the point of hatching, for as previously stated the bees, impatient of delay, will reject some of them.

Snelgrove modification of the Barbeau method.

The following plan, a simplification of Barbeau's method, is used with satisfactory results by the writer. It has the advantage of dispensing with cell-holders which are not too easy to make, and the processes of cutting out and fixing the selected cells are both easier and safer.

The cell-bar is provided with a layer of wax, waxen discs, or waxed cell-holders (Chapter XIV) on its under side and to these the excised cells are to be fixed.

A short length of metal tubing similar to the punch used by Barbeau (p. 260) but of larger internal diameter, viz., $\frac{5}{8}''$, is used, and one of its ends is ground to a cutting edge.

Another similar but longer tube, of external diameter $\frac{1}{2}''$, with both ends blunt, slides easily into the punch, and serves to expel the excised cell without coming into contact with it or its contained larva.

The cell-bar is placed on a table, waxed side or waxen discs upwards.

The comb containing the selected larvae is examined and a patch with very young larvae is cut out. The cells on the *reverse* side of this patch are shortened to about $\frac{1}{8}''$ by a razor. The comb is then laid flat on a table, shaven side downwards, and the punch is used as described on p. 262 to cut out a chosen cell. The punch is large enough to include parts of the six surrounding cells and it is therefore easy to ensure that the selected cell is central, and to avoid damaging it.

An assistant touches the waxen surface or disc with a moderately heated soldering or other iron so as partially to melt it. The smaller steel tube is immediately inserted into the blunt end of the punch and the cell with its surroundings is gently pressed on to the melted surface.

If there is doubt about the adhesion of any of the cells to the bars a little melted wax should be painted round their bases. The extra wax will contribute to the building of the queen-cells.

It will be noted that the cell is merely pushed out of the cutting end of the punch and is not made to traverse its length as in Barbeau's method. In other words the thrust of the expelling tube is in the opposite direction.

The mouth of the selected cell may then be slightly enlarged with the grafting tool. Any larvae in the six surrounding cells must be removed.

The process is so simple and expeditious that a novice can hardly fail to perform it perfectly.

B.H.S. method.

In the "American Bee Journal" of December 1944 B. Hiller describes a method of obtaining select queens. He prepares a cell-starting stock much on the lines of that of the swarm-box described on p. 218. After liberal feeding, which is continued, he places over it a comb of eggs and young larvae taken from his breeding queen, placing it flat and raised above the top bars of the other combs by 1″ pieces of wood. The bees select a number of the cells on the under side and convert them into queen-cells. Hiller considers that "The bees do their own selecting, and they know better than man which cells will make the best queens". He leaves this comb in the cell-building stock for two or three days and then transfers the queen larvae, now floating on "beds" of royal jelly, into artificial queen cups. As he says,—"Anyone can do it, for this bed of jelly is so thick that you can get under the larva without even touching it".

This excellent method can be simplified. If instead of grafting, which involves much work, we combine it with the writer's modification of the Barbeau method, we get the simplest and most effective method of obtaining mounted queen-cells, in the best circumstances, and usually with 100 per cent of acceptances.

The detailed procedure is as follows:—

Comb of eggs and young larvae.

Five days before it is intended to start queen-cells insert a frame with foundation or a fairly new drawn comb, preferably the latter, into the stock from which it is intended to breed. If this frame contains only one or two vertical wires to support the middle of the comb so much the better. Bees often construct queen-cells over the wires which are inconvenient when the cells are being removed. The comb may be shallow or full-sized according to the number of queen-cells desired. If the stock is fed the inserted comb should be filled with eggs on the morning of the fifth day and some of them will have hatched. If this condition is found prepare a cell-starting stock during the same morning. If no eggs are hatching wait another day.

The cell-starting stock.

The stocking of a swarm-box with confined queenless bees (p. 211) is not suitable for this method because of the loss of bees when it is opened. The most suitable plan is that described as Method 24, (p. 218) in which the bees are queenless and broodless, unconfined, and on their own site. The stock should have been fed beforehand and should be provided with syrup in division-board feeders as it will not be convenient to feed it from overhead. It should be provided with as many combs of stores as are available but these should not contain a single egg or larva. Space for the reception of a cell-bearing frame should not be left at this stage but provided later by the removal of a central comb.

Placing the horizontal comb.

When the cell-raising stock has been queenless for from 6 to 8 hours remove the comb of eggs and young larvae from the breeder stock and dislodge the bees (p. 194). Take it, and two *empty frames* of the same size, to the cell-raising stock. Place one empty frame horizontally over the combs of the stock and upon it lay the prepared comb of eggs and young larvae. The empty frame provides space for the construction of queen-cells. Above the comb place the second empty frame to take the weight of the quilts. Enclose all in a lift and leave for two days. At the end of this time a

PLATE XXIX

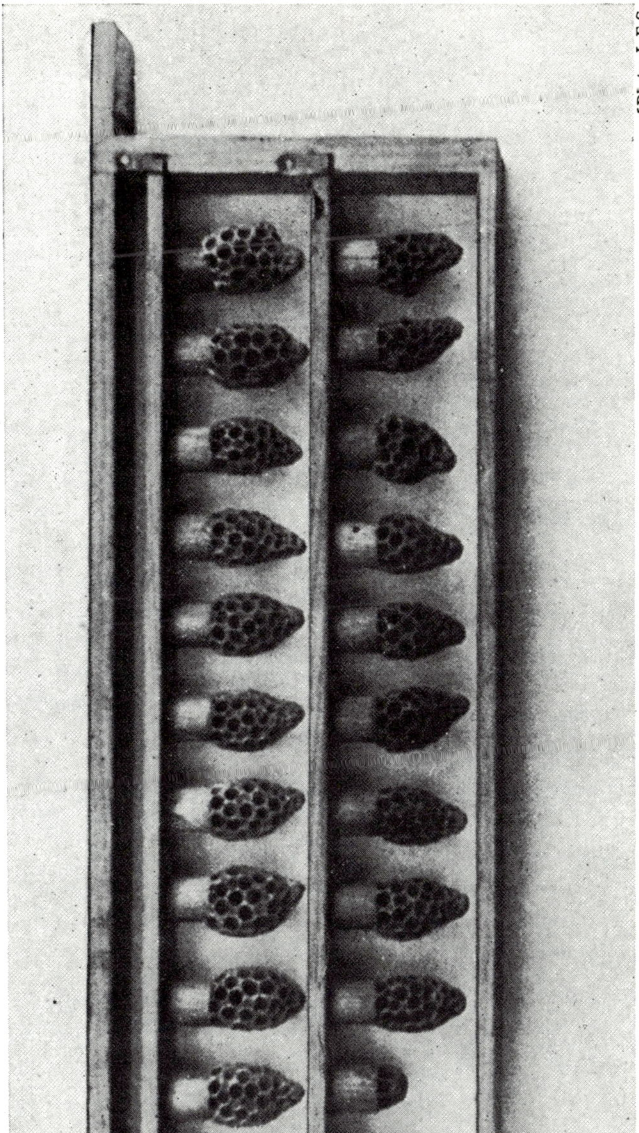

[Photo: L.E.S.

B H.S. METHOD

large number of queen-cells will have been started on the under side of the comb and will be distinguished by the large amount of royal jelly they contain.

Have in readiness one or two frames carrying cell-bars to which are attached cell-bases as described on pp. 232–233.

Remove the comb with queen-cells from the cell-raising stock and after dislodging the bees with a feather take it to a warmed room. If any of the queen-cells are built over wires cut the latter with scissors close to the cells. Proceed to remove them with the punch described on p. 264 and fix them quickly to the cell-holders as described on the same page.

Return the mounted cells to the cell-raising stock for completion. As the bees receive their own queen-cells back after a very short interval they will not reject any of them provided damage and chilling are avoided.

As this method is compounded of those of Barbeau, Hiller, and the writer, it is suggested that it be known as the "B.H.S." method.

CHAPTER XVII

USING ONE STRONG STOCK FOR STARTING AND COMPLETING QUEEN-CELLS

To use a single stock for the starting and completion of queen-cells we may proceed in one of two ways, viz:—

(1) The queen-rearing bees are confined whilst the main stock continues its activities.
(2) The queen-rearing bees are at liberty whilst the remainder of the stock is confined for one and a half to two days.

The first of these methods will now be fully described as being the easier and causing less interference with the ordinary work of the stock.

Method 1.

The steps to be taken are,—
(1) Preparation of a swarm-box.
(2) Stocking the swarm-box with bees.
(3) Reducing the size of the brood-chamber of the main stock.
(4) Rebuilding the main stock with queen below, and supers above an excluder.
(5) Confining the bees in the swarm-box for 6 to 8 hours.
(6) Grafting into artificial queen-cells and insertion of these into the swarm box. Continuing the confinement of the bees for a further 24 hours.
(7) Re-uniting the swarm-box, containing the started queen-cells, with the main stock.

Our aims will be:—

(1) To congest the bees by
- (*a*) Shaking the bees from one super and transferring it temporarily to another stock.
- (*b*) Reducing the capacity of the brood-nest to half its size so that the combined capacity of brood-nest and swarm-box are not greater than that of the original brood-nest.

(2) To discourage the nurse-bees from going down to the queen by leaving only one or two combs of sealed brood in the brood-nest. If this is done there will be no need to place brood in the swarm-box (p. 248).

The following are detailed instructions:—

The stock.

Select a strong stock which is working well in three or more shallow-comb supers above its brood-chamber, or in at least two full brood-comb supers and a brood chamber, such as may be obtained by using the Demaree method (p. 202). All the combs should be well covered with bees and the stock should not be making preparations to swarm. If honey is not coming in feed the stock for three days beforehand with a little honey and water in equal parts.

The swarm-box.

For a swarm box use an extra brood chamber and provide it with two box dummies $3\frac{1}{2}''$ thick, one at each side. The space between these will then be about $10''$ wide, sufficient for

(1) A division-board feeder at each side.

(2) Three combs containing honey and pollen, but no brood.
(3) A central space for a frame carrying cell bars.

A framed screen of perforated zinc forms the base of the box—to which it is attached by carpet staples—and a shallow wooden cover fitted with two half-length withdrawable slides (Pl. XX) to allow of the insertion of the cell-carrying frame, is provided. Two $1\frac{1}{2}''$ holes, one on each side of the central slide, are contrived in the cover to be used for extra ventilation when necessary, or for additional feeding whilst the confined bees are engaged in constructing queen-cells. A pivoted piece of perforated zinc covers or uncovers each hole as necessary. A lift, to accommodate feeders and quilts, may be used *above* the cover.

To stock the swarm-box.

At about ten o'clock in the morning take the swarm-box to the selected strong stock, having previously furnished it with box dummies, two combs of honey, one comb with pollen, and two division-board feeders, leaving the central space empty for the reception of the cell-bar frame. Remove the supers from the stock and place them on one side. Remove the comb of bees on which the queen is found and place it temporarily in a nucleus hive.

With or without the funnel (p. 214) shake the bees from the remaining brood-chamber-combs into the swarm-box. At this stage the cell-carrying frame may be inserted into the central space in order that the bees may fashion the artificial cells to their liking before they are grafted, but it is much more convenient to get this done previously in the super of the same or another stock.

Close the box with the lid and provide the latter with two feeders—one containing honey and water, or syrup, and the other if possible, with a jar of wet cappings fresh from the extractor (p. 212).

Take the box, provided with a lift surmounting the cover to accommodate feeders and quilts, to a cool shady place where it is to remain for about 6 hours. Raise the screened base an inch or two from the ground to give the confined bees ample ventilation.

Rebuilding the stock.

Replace the comb of bees with the queen in the original brood-box. Add two of the shaken combs containing sealed brood and two broodless combs. Insert two other box dummies, one on each side.

(Place all remaining brood-chamber-combs temporarily over an excluder on another stock).

Restore supers above the excluder.

The stock will now have a diminished brood-nest to compensate for the reduced population and the queen will have laying room for the period during which queen-cells will be maturing.

Grafting.

Towards evening remove the cell-carrying frame (p. 194) and take it to the grafting room. Take the comb containing one-day-old larvae from the breeding stock and carry it to the grafting room in a warmed nucleus box (p. 240). Graft the cell-cups as directed on page 241 and insert the cell-carrying frame into the swarm-box.

Use of carbolic cloth.

We must here refer to the difficulty of opening a swarm-box containing imprisoned bees. If the slides are withdrawn for the purpose of inserting or withdrawing the cell-bearing frame large numbers of bees will escape, and the smoker is not effective in preventing them from doing so. The carbolic cloth (20, p. *101–2*) however is most effective and does not in the least affect the success of the operation. The cloth is spread over the lid of the box and the slides withdrawn from underneath it. The bees retreat from it to the innermost parts of the box and after half a minute the cloth may be removed and the frame inserted or withdrawn for inspection with scarcely any loss of bees. Similarly the cloth is useful when removing the screen from the bottom of the swarm box, as referred to in the next paragraph. The box is placed over the outspread cloth for half a minute, after which it may be lifted off the screen without loss of bees.

The cell-completing stock.

After an interval of from 24 to 30 hours the bees in the swarm-box will have made good progress with the queen-cells they have accepted and the whole stock may be re-established. The swarm-box is lifted off its ventilating screen and placed over an excluder which has been previously placed above the supers. The partially developed queen-cells will be completed by the whole stock in due course.

It will be noted that the total capacity of the hive has not been increased, the queen being in a half-sized brood-nest at the bottom of the hive. The queen-cells separated from her by two excluders and intervening supers are completed by crowded bees in the presence of ample stores. The queen-cells illustrated in Pl. XXIX, were completed in this way.

Method 2.

Alternatively, the swarm-box, when equipped and stocked, may be placed on the original hive floor and the bees allowed free flight. In this case rather less bees should be shaken into it as it will become very strong through the accretion of flying bees. If liberally fed a somewhat better acceptance may be expected than if the bees are confined.

During the 24 to 30 hours in which the queen-cells are being started the whole of the parent stock, built up as described on page 272 must be confined over a screen and kept in the shade. It is then restored to its former position, liberated, and the swarm-box placed over it and above the excluder.

Uniting swarm-box and stock.

When uniting a stocked swarm-box to a main stock the use of the second excluder is desirable but not essential. It serves to increase the feeling of queenlessness of the bees completing the queen-cells and to discourage them from breaking down a proportion of these when nearing maturity, which they occasionally do in bad weather or in the late season.

If the bees of the swarm-box are kept apart from the main stock for more than two days, some fighting may ensue after uniting. The writer guards against this possibility by always uniting with a sheet of newspaper. Whilst not essential, this ensures not only that there will be no fighting but also that there will be no sudden abandonment of the queen-cells by the nurse-bees which at first tend to go down to the queen below. The union is peaceably effected within an hour or two and the bees gradually assort themselves according to the duties they have to perform.

Raising queens over a screen.

It is possible to get queen-cells started in a honey-super which is temporarily separated from the main stock by a screen of perforated zinc. This however entails the necessity of adding a considerable force of young bees taken from brood combs to the bees in the super. If this is not done a bad result like that illustrated in Pl. IX is to be expected. Pollen, and brood in advanced stages only, must also be placed in the super.

Grafted queen-cells are inserted in the super after the bees have been confined for a few hours, and a day (or two days) later the screen is removed so that the cells may be completed by the whole stock.

This method is not generally to be recommended however for the proportion of acceptances is always smaller than when a state of complete orphanage is imposed. Moreover it involves all the preparations needed for a separate swarm-box.

If brood is not provided in the super most of the nurse-bees return quickly to the queen and brood-nest below when the screen is removed, and in consequence the young queens to be reared will be insufficiently nourished.

CHAPTER XVIII

FORMATION, CARE, AND DISPOSAL OF NUCLEI

A SMALL colony of bees capable of growing into a full-sized stock is usually referred to as a "nucleus". Strictly speaking it should contain stored combs, bees, brood, and a queen, but for convenience we shall call it a nucleus whether it has a queen or not. Every young queen, in order to become mated, must be able to fly freely from a stock, which may be large or small, and to which she may return to commence egg-laying.

The queen-breeder needs at any one time, as many nuclei as he has of newly mated queens, virgin queens, and mature queen-cells ready for distribution.

The abstraction of the necessary combs and bees from strong stocks for the provision of these nuclei necessarily limits the honey harvest, but if done in early summer it tends to delay swarming. After the honey harvest there are usually plenty of bees and combs available for this purpose.

Nuclei formed at about swarming time are likely to remain strong and to collect much of their own food. Those made after the honey harvest need more or less constant feeding.

Nucleus hives.

It is not necessary here to describe all the types of nucleus hives in use. They vary principally in devices for feeding, or in arrangements for housing nuclei side by side for the sake of extra warmth. It will be sufficient if we consider essential features of typical hives used by the writer and illustrated on p. 307.

Whilst we should observe reasonable economy in bees

when forming nuclei we must ensure that there are sufficient to maintain the natural hive temperature, and therefore there must be at least one seam of bees between two combs. A one comb nucleus is liable to chilling in cool weather unless it is sandwiched between other similar nuclei in a long divided hive. Three combs well covered with bees provide two interior seams of bees and make a stock which can be maintained throughout the season, and which will develop sufficient heat to "hatch" a queen-cell and to rear the first brood of a young queen. Additional lateral space will be needed to accommodate a fourth comb, or a division board feeder when needed, and also to allow for temporary wider spacing of two of the combs when a ripe queen-cell, with or without its protector (p. 299), is to be inserted. Now four frames with metal ends occupy 6″ of lateral space and if we allow another 1″ for wider spacing we conclude that the internal width of a four-comb nucleus should be 7″. Other dimensions are those of an ordinary brood box.

The floor should be detachable but easily fixed for transit. An entrance of 3″ × $\frac{3}{8}$″ is suitable for warm weather but should be reducible, by means of a metal slide or otherwise, in order to guard against robbers in the later part of the season.

A detachable lift sufficiently deep to contain a 2lb. lever-top tin feeder and necessary quilts is a great convenience. In cold weather it is sometimes necessary to feed without opening the hives to replenish the division-board feeder, but this cannot be done in the case of nucleus hives without lifts. A water-proofed cover is necessary and a strong handle of cord or wire for carrying the hive is useful.

In very hot weather life inside an unshaded nucleus hive becomes almost intolerable. To provide for additional ventilation when needed a hole 1″ in diameter should be bored in the upper portion of the back of the hive and this covered on the inside with perforated zinc. The hole is stopped with a cork on the outside when extra ventilation is not needed.

Shallow comb nucleus hives.

For some years the writer has worked a number of stocks on shallow combs only. There are many advantages and one or two disadvantages in this which will readily occur to the experienced beekeeper. For queen-rearing there is the great advantage that efficient nuclei can be formed with fewer bees and that they are lighter and more easily examined than those formed with standard brood combs. Simply made hives of this kind are illustrated in Pl. XXXIV. The interior width is $8\frac{1}{2}''$ which permits of an initial nucleus of three combs with a division-board feeder, to be built up to one of five combs when the feeder is withdrawn. These hives have proved entirely satisfactory in use.

Some beekeepers winter their stocks by supplementing the food of each with a shallow-comb super containing sealed honey. This super constitutes an enlargement of the brood-nest in spring, and the shallow combs, when occupied by brood, are available for shallow-comb nuclei.

Nuclei in divided brood-boxes.

By means of plywood division boards sliding in grooves any brood-box or shallow-frame super of standard dimensions may be divided into bee-tight compartments for separate nuclei. For example the "W.B.C." brood-box will accommodate three nuclei of three combs each but provide little additional space for lateral movement of the frames. The square 11-frame brood-box may be partitioned for one four-framed and two three-framed colonies, but in use the frames are sometimes not easily withdrawable; three equal compartments, each for three combs, with light dummies to take up surplus space, are to be preferred.

The floor of a divided box may conveniently be

similar to a "Swarm control" board (78, p. *28*) which is easily fixed to the box by four carpet staples. Wooden strips fill the spaces under the plywood division boards, and three wedge openings, one at the front and one at each side, serve as entrances. Small differently coloured alighting boards are affixed under each opening.

Three quilts are needed—one for each compartment. One of the difficulties experienced in manipulating nuclei side by side is that of preventing the bees from one nucleus from entering a neighbouring one when quilts are raised. To obviate this the writer provides three quilts of canvas. The two to cover the side compartments are cut slightly large and each is fixed by drawing pins along the length of the plywood division board. The quilt to cover the middle compartment is cut three or four inches wider so as to overlap the other two, and is not fixed. Any compartment can then be examined without opening the others. The greatest care is taken that bees cannot pass from one compartment to another.

Each quilt is provided with a feed-hole and the whole box is surmounted by a plinthed lift to accommodate feeders and quilts. The whole box with the fittings can then be placed above the supers of an ordinary stock, or placed on a separate stand.

Long hives are sometimes divided in the manner described above. Each compartment should have room for three combs, and if possible a dummy, the removal of which will provide space for lateral movement, or for an additional comb. Smaller nuclei are not recommended.

A long hive built for 22 combs can be divided into six 3-comb compartments. Its entrances should be situated as follows:

No. 1. At end, floor level.
No. 2. ,, front, 4″ or more above floor level.
No. 3. ,, back, floor level.

No. 4. At front, floor level.
No. 5. ,, back, 4″ or more above floor level.
No. 6. ,, end, floor level.

With this arrangement all the entrances are as far apart as possible and if they are provided with small differently coloured alighting boards (blue, yellow, or white with black stripes), returning young queens are not likely to enter the wrong ones.

Nuclei in divided hives are not generally to be recommended. Although they have the benefit of the mutual conservation of warmth, and some ingenious arrangements have been devised whereby two or more nuclei can feed from the same receptacle, they are less mobile and more difficult to stock than single nucleus hives. To explain this last point:—

Suppose we form half a dozen queenless nuclei on the same day and bring them home from an out-apiary to await the insertion of mature queen-cells. We set them down in the garden near to one another, and liberate the bees. They fly out, excited and distressed, begin to take their bearings, and see not one but half a dozen new and unattractive homes before them. By the next day they will have congregated into one or two of the hives leaving the others almost if not quite abandoned. This is what sometimes happens if a number of nuclei are put into a divided hive at one time, and especially if all the nuclei have been made from the same stock. We avoid this tendency to join up in the case of single hives by placing them some distance apart. In the case of the divided hive we either stock the compartments at separate times, or all at the same time with nuclei already possessing virgin queens and derived from different stocks.

When two or more nuclei have access to the same feeder the stronger and perhaps the less needy takes the greater share of the food.

Baby nuclei.

Some writers, notably Pratt of U.S.A. (65, p. *1*) record success in the use of miniature hives for accommodating tiny stocks of bees with virgin queens, pending the mating of the latter. In some cases as little as half a pint of bees is considered sufficient. Small frames of sizes one-fourth or one-sixth the size of a standard frame are fitted into the latter, provided with foundation, and inserted into a strong broodnest until furnished with stores and brood. When removed a small top bar is fitted to each and two or three without bees are inserted into a "baby" nucleus hive made to take them. About half a pint of bees are inserted and confined for a few hours. They are then given a ripe queen-cell, taken to an out apiary, and liberated after dark. The hives may be placed in any shaded position, each being provided with a feeder.

The writer who made extensive trials with "baby nuclei" some years ago hung them in apple trees. Although a small proportion of queens in these hives were successfully mated and began to lay, by far the greater number led their bees off as swarms when they issued for their mating flights. Neither old nor young brood would encourage them to stay and they probably absconded for some of the reasons given in the next paragraph.

Stable conditions for a nucleus stock.

The following conditions are requisite for the stability of a nucleus stock.

(1) Sufficient bees to maintain the normal hive temperature and to maintain adequate ventilation.
(2) Presence of sufficient young bees for domestic duties, and of older ones for foraging, not only for nectar and pollen, but also for water, which is equally essential.

(3) Stores of honey and pollen.
(4) Brood,—mostly old and sealed if the stock contains a virgin queen. Brood of any age may be present when the queen is fertile.
(5) Shade during the heat of the day.

If any of these conditions are absent the little stock will be unbalanced, and not taking kindly to its new circumstances will most likely take the first opportunity —afforded by a hot day and the flight of a young queen— to abscond.

Wedmore (94, p. *69*) observes that if a nucleus comprises only very old bees these are apt to swarm with the queen when she flies to mate. To this the writer would add that in the case of baby nuclei, absconding will frequently take place even when all the bees are young and have not previously learnt to fly.

Absconding swarms are a source of frequent loss to queen-breeders. It is easy to realise that a "Baby" nucleus can seldom be balanced and may therefore be expected to abscond.

In general it is found that the stronger and more prosperous a nucleus or stock the more likely is its young queen to become safely mated.

To summarise,—

We should prefer separate nucleus hives each accommodating at least three combs with room for some lateral movement, and a feeding division board or fourth comb; accommodating standard frames, whether deep or shallow; adapted for feeding; and easily closed and moved.

Formation of nuclei.

Reference has already been made to the formation of nuclei in particular circumstances in Chapter VIII. We may repeat here that it is a great advantage if newly formed nuclei can be moved to a distance (e.g., to a friend's garden more than a mile away, or to an isolated queen-rearing apiary) so as to avoid loss of flying bees; if however the nuclei are to be retained in the apiary where they are formed extra bees must be added to them to compensate for this loss.

To obtain nuclei for queen-rearing we may adopt either of two courses:—

(1) We may sacrifice whole stocks, making three or four nuclei from each.
(2) We may take a nucleus from each strong stock in an apiary just before an expected nectar-flow.

The latter course is preferable for a vigorous stock soon recovers from its loss at this time of year especially if shallow combs are used. The abstraction of a nucleus has the effect of delaying swarming, and may in rare cases prevent it. We now give—

(1) The orthodox and commonly recommended method of making a queenless nucleus stock.
(2) The writer's method, which is more expeditious and satisfactory.

Usual method.

Following the directions as given by Cowan (20, p. *132*):—"Remove from a populous stock two or three combs one of which should contain honey and pollen and the others brood. . . . The combs are removed with the adhering bees, care being taken that the queen is not with them. Place these in the nucleus hive. As all the old bees will return to their hive shake or brush the young ones from one or two other combs into the nucleus so as to introduce about a quart of bees, which will stay and be sufficient to keep the nucleus at a proper temperature. Insert empty combs or frames of comb-foundation into the hive from which the full combs have been removed."

There are two main objections to this procedure, viz.:—

(1) It is necessary to find and temporarily remove the queen which may take some considerable time.
(2) Drones will be included in the nucleus. If these are not of the desired variety or strain they will vitiate attempts at selection. Droneless nuclei are essential for the queen-mating apiary.

By using the following method we may avoid these disadvantages:—

Snelgrove method.

Take an excluder and an empty brood-box together with three frames of foundation to a strong stock which is not preparing to swarm. Remove the supers and excluder, if any. Look through the brood nest and select:—

(1) one comb of honey and pollen.
(2) two combs of brood, mainly sealed and without drone comb.

Place these temporarily into the extra brood-box and put this over the brood nest. Gently shake the bees from the three selected combs down into the brood-nest. If the queen is on one of them she will fall with the bees. Now remove the extra brood box, with the three shaken combs, to one side.

Insert the three frames of foundation into the brood nest and re-build the hive as follows:—

Brood nest, excluder, supers, second excluder, extra brood box with the three shaken combs, quilts. After 15 minutes, or at any time later, raise the quilts. The three combs will now be thickly occupied by bees (without drones) which have come up through the excluder. Place them in a nucleus hive and close this at once. Remove the spare brood-box and excluder and restore the quilts.

Nuclei so prepared are to be removed to another apiary as it is not necessary for them to lose flying bees.

The whole process takes little time. About half a dozen nuclei can be formed and hived within an hour if first they are all prepared, and then collected in rotation.

Sometimes the bees will be unnecessarily thick on the selected combs. If so some can be gently shaken off. They will comprise a large proportion of young bees which will ensure the prosperity and continuance of the nucleus.

Should a few cells of drone brood be found on the selected combs they should be destroyed with a penknife.

Nuclei made in this way prove to be excellent. They do not abscond, and maintain their strength throughout the season if successive young queens are allowed to lay in them until tested (p. 312).

We need hardly add that nuclei should not be made in this way from stocks preparing to swarm. The operator must therefore watch carefully for queen-cells when first looking through the stock.

Care of nuclei.

When the nuclei have been formed they should be closed,—with ventilation—removed to positions in the apiary where queen-cells have been raised for them, and liberated. If they have to be moved subsequently to a mating station this should not be done until the young queens have emerged. The distribution of mature queen-cells to them is considered in the next chapter.

As a nucleus is unlikely to have many foraging bees it must be fed to keep it in good heart. Both honey and pollen are necessary for the nutrition and full development of the young queen during the period between her emergence and mating.

Reference has already been made to alternative methods of feeding, viz. by way of the feed hole in the quilts and by the division-board feeder. The latter is similar in shape and size to the frame in use, and may be fitted with a tin receptacle for syrup,—replenished through a hole bored through the top bar. When this is hung in the hive the bees have access to the syrup from the top of the receptacle, which does not reach to the top bar. They empty it quickly, excitement and temperature increasing as they do so. It is specially suitable for use in the swarm-box (p. 207).

A division-board feeder for syrup may be made of a frame and 3-ply sides, provided that the joints be sealed by varnish or mastic. In this case the tin receptacle is not needed.

The writer considers that if there is a tendency for the bees to abscond with the young queen, rapid feeding by means of syrup promotes it. For use with mating nuclei therefore he fills a feeder with candy which is used up very slowly and without causing excitement. One filling is sufficient to provide food for a long time,—at least until the queen has begun to develop a brood-nest.

This feeder is illustrated in Pl. XXX. It consists of an ordinary shallow frame one side of which is completely covered by a fixed sheet of ply-wood. For the other side a movable panel of ply-wood covers the side except a ¾″ space at the top. This panel is held in position by two small internal fixed strips fitting under a strip of ply-wood extending along the lowest part of the frame, and by two wooden buttons at the top. The construction is simple and is shown in the illustration. The interior is coated with knotting or varnish to make it non-absorbent.

Laid flat on a table it is filled with candy in the ordinary way, and hung in the nucleus hive—at one side—when the nucleus is formed. As it holds 2½lbs. of candy it may not need replenishing. If all goes well

PLATE XXX

[*Photo: L.E.S.*

Division-board Feeders Provided with Candy for Use in Nuclei the nucleus will begin to store honey and then the feeding division-board should be removed and a frame of foundation substituted for it, but in continued bad weather further feeding may be necessary.

Congestion in a nucleus.

Congestion may occur in a nucleus,—

(1) If there is a heavy nectar flow.
(2) If a fertile young queen is not removed when she has filled the combs with brood.

In each case foundation should be substituted for full combs—the latter going to strengthen weaker nuclei, or to contribute to the formation of new ones. If a congested nucleus is neglected, even as late as August, the bees will swarm and the queen may be lost.

On the other hand a nucleus of which the queen fails to mate, or is lost, and which consequently is without brood for a long time, may become too weak for further use.

To strengthen a nucleus.

Should a nucleus need to be strengthened a comb of emerging brood should be given to it. Failing this the plan suggested by Wedmore (94, p. 59) may be used, viz:—

Shake one or more brood combs of bees near the front of the nucleus. The older bees will return to their own hive if this is in the same apiary and the young ones will enter the nucleus where they will be well received.

Disposal of nuclei at the end of the season.

If weather is favourable and drones have been preserved September is quite a good month for the mating of queens. During the last week of that month however all queen-rearing should be discontinued and the nuclei disposed of in any of the following ways:—

(1) Strong ones may be wintered separately.
(2) They may be added to stocks needing requeening.
(3) They may be united to form new stocks.

Wintering nuclei.

A four-comb nucleus containing a young queen and at least two combs well filled with brood will winter quite well if fed with 10 lbs. of thick syrup, warmly covered with quilts, and kept in a place sheltered from cold winds. It will need to be fed again in late spring, and when strong, transferred to a larger hive.

Several nuclei can be wintered side by side in a large packing case through the sides of which suitable entrance holes are bored. The hives should be covered with hay or straw and a waterproof roof provided. The extra queens so wintered will be of considerable value in the spring.

Requeening with a nucleus.

Remove the old queen from the stock towards evening. Place a sheet of unbroken newspaper over the combs so as to cover them entirely If late in the season and the weather is cold prick a few holes in the paper with a fine wire nail. Place an empty brood box containing two division boards spaced apart over the paper and transfer the nucleus stock to it. Replace the quilts. During the night the bees nibble away the paper and unite peaceably. A week later put the nucleus combs down into the brood box, withdrawing others to make room for them.

Uniting nuclei.

Three fairly good nuclei, two at least containing brood can be united to form a strong stock, headed by one of the queens.

If the union is to be effected in the apiary in which the bees have been accustomed to fly it will be necessary gradually to bring them near together until they stand side by side. The united stock will then occupy the place of the middle one.

The old-fashioned method of uniting would have been to remove two of the queens (needed elsewhere) and to sprinkle flour, or spray scented water over all the bees of the three lots before putting them together in the new hive. But this was laborious and troublesome business, and has long been superseded by the newspaper method.

If the weather is not cold the three nuclei may be joined by the newspaper method as described above, the one with the chosen queen at the bottom, and the others above it, separated by sheets of newspaper. Should any one of the stocks have been queenless on account of the earlier removal of its queen, it is advisable although not generally necessary, to destroy the queen-cells before uniting. The combs of the united stock should be rearranged a week later. This method however it unnecessarily cumbrous.

In mild weather three nuclei which are standing side by side and which have been previously fed may be united as follows:—

Remove the queens. Towards evening take six of the combs with bees, two from each nucleus, and place them in the new hive at wide intervals. Open out the remaining combs in the nuclei and expose all to the air for five minutes. Bees and combs will gradually lose their respective hive odours and the combs may then be placed, as three separate lots side by side—not alternated—in the new hive. If no fighting is observed cover with quilts, and before nightfall introduce one of the queens by the water method (79, p. *191*), or if preferred by caging. This method is obviously not suitable for cold weather.

The easiest and most expeditious method of joining nuclei is that used by the writer. It involves the transport of the bees from one apiary to another or to a reasonable distance, and is effective in any circumstances. Moreover it is not necessary that the nuclei to be united be standing near together.

The procedure is as follows:—

Take an empty brood-box and two framed perforated zinc screens, one already fixed by staples to the bottom of the box and the other ready for fixing over the top. Remove all three queens—reserving the chosen one in a cage. Towards evening collect the three nuclei from wherever they happen to be. Place all three stocks, without ceremony, side by side in the brood-box and fasten them in with the upper screen. Put them in the car or on a truck and take them to a distant place. The shaking they get on the journey unites them perfectly. On arrival remove the lower screen and put the stock on a hive floor. Remove the top screen and substitute for it a quilt. After giving the bees an hour or two to settle down introduce the queen at the feed hole by the water method for which a tin of warm water is kept available.

Nuclei kept in a divided hive can easily be united in the following way:—

Assuming there are three nuclei:—Remove two of the queens during one evening. About 24 hours later unfasten and raise the quilts near the ends of the frames and place across the end of each of the division boards or partitions a short strip of wood of about ½″ square section. When the quilts are lightly replaced, these strips of wood will afford small passages through which the bees can pass from one compartment to another. As union will be gradual and at night there will be no fighting. A few days later the partitions must be withdrawn, the combs properly spaced, and only one entrance left for the combined stock.

To avoid loss of field bees nuclei should be united on cool or wet days when bees are not flying. Queens should be introduced to united stocks towards dusk when all the bees are at home.

CHAPTER XIX

DISTRIBUTION OF MATURE QUEEN-CELLS

HAVING dealt with the starting of queen-cells in the swarm-box and their completion in the cell-building stock we must now consider the procedure necessary to ensure that each young queen emerges from her cell either into a stock or nucleus specially prepared to receive her, or into a nursery cage. The procedure applies equally to natural and artificially raised cells.

It is first expedient to consider the circumstances under which queen-cells are well received.

As already stated a stock will begin to raise queen-cells about 12 hours after its queen has been removed. During this period it will not be disposed to accept either a queen or a queen-cell from another hive although it may be caused to do so. After an interval of twenty-four hours a number of larvae will be receiving royal jelly and queen-cells will be in course of construction over them. At this stage the bees will not object to the introduction of one or more cells in a similar (i.e., unsealed) stage, but they will usually tear down a sealed one. This point is important to the queen-breeder especially in view of the fact that some writers recommend, without qualification, the introduction of a "ripe" queen-cell 24 hours after the removal of the queen. On the third or fourth day however when their own queen-cells are nearing the time for sealing they have much less objection to the reception of a sealed cell and usually accept it. When some or all of their own are sealed there is little or no danger to an imported one.

Generally we may say that when bees have eggs or young larvae from which they can raise their own queens they are not so well disposed to receive a sealed cell from another stock as they are when they have no means of raising their own.

Similarly in queen-rearing a batch of artificial grafted cells is likely to be wholly or partially neglected if it is possible for the bees to raise a single queen-cell on their own combs.

To introduce an unprotected sealed cell therefore we should wait until the end of the third day, or better still, until the seventh or eighth day, to make sure of acceptance.

Unsealed queen-cells are almost always acceptable to queenless bees but they should not be introduced to nuclei for these are unsuitable for the completion of nutrition and incubation.

Queenless bees are unlikely to tear down sealed queen-cells presented to them if a nectar flow is on, or if they are being heavily fed. Jay Smith (77, p. *57*) makes a great point of this. He relates how he was at one time puzzled about his severe losses through nuclei repeatedly tearing down their cells. Reflecting that sealed cells could always be safely transferred from one cell-completing stock to another he realised that the main difference in condition between the cell-completing stock and his nuclei was that the former was being liberally fed whilst his nuclei were not. His losses ceased when he fed the nuclei.

Jay Smith (77, p. *63*) apparently inserts queen-cells into nuclei on the day when these are formed but he speaks of losses, and states that unless these nuclei are abundantly fed they will tear down nearly all the cells.

He further states that during a honey-flow or with heavy feeding a mature queen-cell can be given to a stock immediately after the removal of its queen (77 p. *64*). Herrod-Hempsall (33, I, p. *634*) however says that this cannot be safely done during a nectar flow, but that it is quite safe when the nectar flow is over,—

an interesting example of how great minds may differ.

Bees that have been queenless *and* broodless for a few hours will accept an unprotected queen-cell, or a queen, whether virgin or fertile.

The saving of time is important to the queen-breeder. He does not wish to keep his nuclei waiting till the third day or later before introducing his queen-cells, and he wishes to avoid all risk to the cells he has taken so much trouble to obtain.

His best course is to use cell-protectors in all cases. Now Jay Smith (77, p. *52*) does not favour cell-protectors. He considers the bees cannot properly care for the imprisoned cell and that if they are inclined to tear it down, and are prevented from doing so, they will kill the virgin queen when she emerges. With all respect the writer is bound to say that he has never had this experience. His occasional losses, never more than one per cent when protectors were used, have been due to the deaths of queens in their cells.

The times of introduction have much to do with success. We can imagine bees being hostile to the inmates of cells given to them before they have felt the desire for a successor to their own queen, for these circumstances are inconsistent with nature; and we might expect the hostility to persist during the short interval before the queen emerges. Pritchard (67, p. *13*) recommends the use of protectors. The safe rule therefore is, to leave a nucleus queenless for a day and then to introduce a protected cell.

Cell-protectors.

The spiral cell-protector illustrated on page 299 is most commonly used and is both cheap and efficient. Those usually sold by manufacturers, whilst suitable for unmounted cells, are not sufficiently large for some queen-cells cut from the comb or for those mounted on wooden cell-holders. Suitable ones are made for the writer by Mr. J. Richards who gives the following directions:—

A convenient length of wooden rod (e.g., part of a broom handle) of $1\frac{1}{8}''$ diameter is tapered to a point so that the tapered surface is $2\frac{1}{4}''$ long and bulging somewhat at the middle. The small end is cut away at a point where the diameter is $\frac{3}{8}''$ and a shallow saw-cut is made with a tenon-saw across the diameter of the new end sufficient to hold a wire.

The protector is made of galvanised binding wire, of 22 gauge or thereabouts, which is specially suitable and easily obtainable.

Taking a five-feet length, insert one end into the saw-cut. This will prevent the protector from slipping round the rod as it is being made. Wind the wire closely around the rod until a spiral of $1\frac{3}{4}''$ in depth is made. Remove the protector and bend the remaining length of the wire upwards for $2''$, and then horizontally in opposite directions to serve as a support for suspension of the protector between two combs. Cut off the end which was inserted in the saw cut. The opening at the small end of the protector made in this way will be $\frac{7}{16}''$ in diameter which will allow the tip of a queen-cell to protrude from it about $\frac{1}{8}''$. The body of the protector is large enough to take queen-cells cut from the comb, or mounted on any of the cell-holders, except that of Maisonneuve, illustrated in Plate XXIV. It will appear to be large for queen-cells not mounted on wooden holders but that does not matter.

The protector is completed by a small square of zinc or tin which is lodged between the upper turns of the spiral to form a roof. This prevents the bees from entering the protector from above.

The mature cell is inserted into the protector until its tip protrudes $\frac{1}{8}''$ or slightly more through the narrow opening at the bottom. The lower part of the cell is extremely tough because of the cocoon inside and the bees cannot bite through it to attack the queen. They can however easily bite through the waxen sides of the cell above the cocoon and it is to prevent this that the protector is used. After a few hours they lose the desire to attack the queen.

Alternately a cell-protector may be made of a cone-shaped piece of wire cloth or perforated zinc.

The Maisonneuve cell-holder (Pl. XXIV, 5) is itself an efficient protector for cells which are sufficiently slender to be inserted into it.

The weather may be chilly at the time when queen-cells must be distributed. As the pupae are still delicate they are easily killed by exposure to low temperatures, and the metal of the protectors rapidly abstracts heat from thc queen-cells, especially in cold weather. We should avoid this danger either by placing the protector in the hive to be warmed beforehand and taking the queen-cell to it in a warmed cardboard box, or what is far more satisfactory, by using the warmed and fitted nucleus box described on page 240.

Let us assume that we wish to distribute 10 queen-cells to as many nuclei without any possibility of their being chilled.

(1) *Preliminary preparation of nuclei.*

The nuclei should have been queenless for 24 hours or longer, and are kept in the vicinity of the cell-completing stock (Chapter XV) where they are to remain until the queens emerge before being taken to the mating apiary.

First visit each nucleus. Break down any queen-cells that may be found. Separate the combs so as to leave a space between two combs of brood sufficiently wide to receive a cell-protector. Replenish the feeder and close the hive.

(2) *Protecting the queen-cells.*

Take the warmed nucleus box to the cell-completing stock. Remove the frame carrying the queen-cells,

PLATE XXXI

[Photo: L.E.S.

DISLODGING BEES FROM MATURED QUEEN-CELLS. FINISHING WITH A FEATHER dislodge the bees with a feather, and insert the frame into the warmed box. Carry this to a warm room and put each cell into a protector. Insert the metal covers in the protectors and the latter into the board cut to support them in the warmed box (Pl. XXXII). Close the lid of the box and take it round to the nuclei.

In warm weather the use of a warmed box may be dispensed with provided the distribution of the cells can be accomplished within a short time.

(3) *Distributing the queen-cells.*

Suspend a cell in each nucleus, taking care that the metal cover is still in place and that the warm box is kept closed during the intervals.

A few minutes will suffice to introduce all the cells which will at no time have been exposed to a harmful temperature. Moreover they will have been kept the right way up and not laid on their sides as is sometimes depicted in illustrations. Both Pritchard (67, p. *13*) and Jay Smith (77, p. *53*) attach importance to this point especially when cells are handled earlier than the 10th day, the latter quoting Snodgrass as stating that when an immature queen is laid on her side for some time her circulation is impeded and may stop,—but that if the queen is within 24 hours of emerging no harm will result from her being placed on her side for a short time.

Shock or jarring vibrations may be injurious to young queens in their cells and it is therefore inadvisable to transport nuclei to a distance immediately after the insertion of queen-cells. It is quite safe to do this after the queens have emerged.

Nursery cages for surplus queen-cells.

If we have more mature queen-cells than nuclei to accommodate them we may put the surplus cells into nursery cages (Pl. XXXIII). These in turn are placed in the supers of queen-right stocks until the queens have

PLATE XXXII

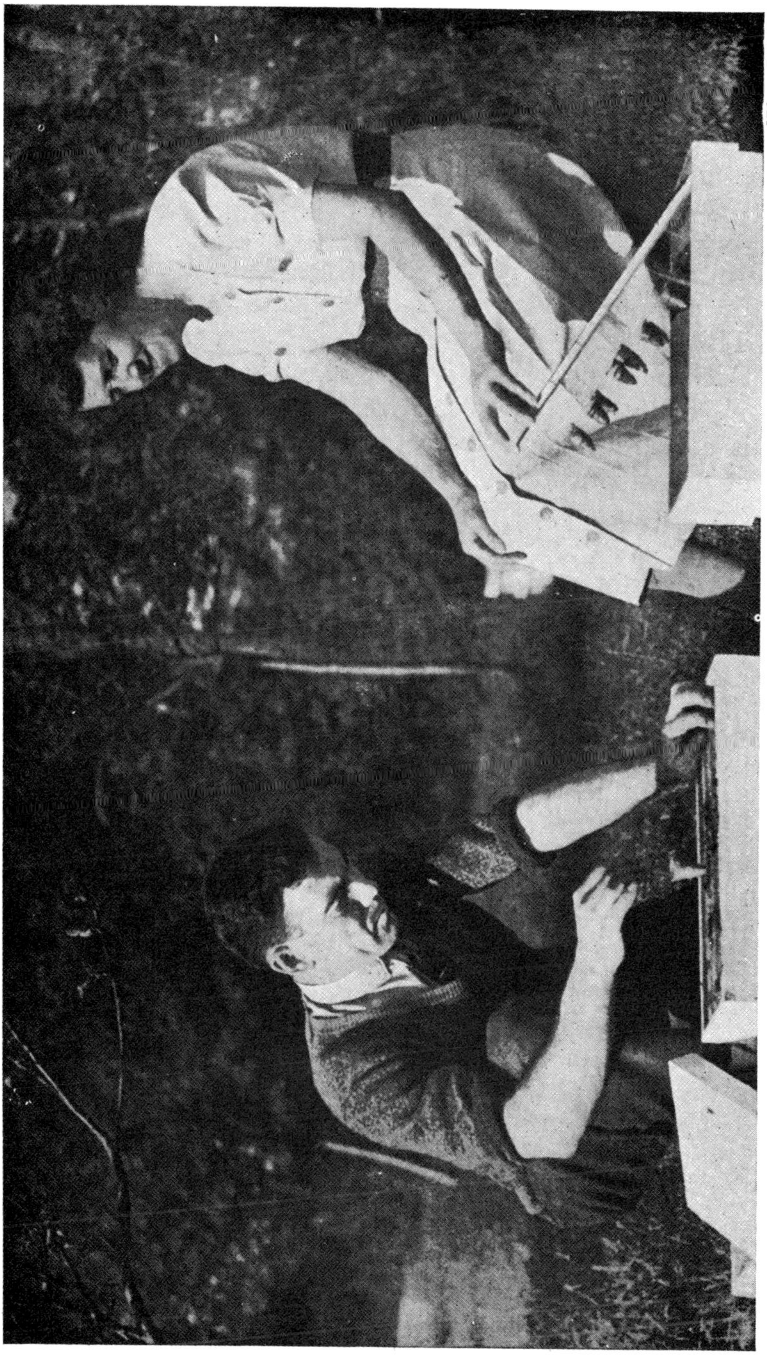

[Photo: L.E.S.

DISTRIBUTING PROTECTED QUEEN-CELLS TO NUCLEI

emerged and are two or three days old. As these cages are entirely closed the bees cannot injure the cells or the newly-emerged queens and the latter find an immediate provision of food in the cages which sustains them until they are liberated. The following description of nursery cages is taken from the writer's "Introduction of Queen Bees," page 176.

"Many forms of nursery cages have been invented, the simplest being that originally made by Mr. H. Alley. This consists of a block of wood $2\frac{1}{2}''$ square and $1''$ thick. Through the middle of the large faces a hole of $1\frac{1}{2}''$ diameter is bored to form the queen chamber. This is covered on both sides by wire cloth. Through one of the small faces, viz., that to be at the top when the cage is in use, two holes are bored into the queen chamber, one of $\frac{3}{4}''$ diameter, with finger depressions, to accommodate a queen-cell, and the other of $\frac{3}{8}''$ diameter to hold candy. The candy hole is to be covered with a small piece of tin which may be turned on a nail when it is desired to allow the bees to liberate the queen. The queen-cell is suspended in the larger hole by means of the rim of its wooden cell-cup, or, if a natural one, by a cork or thin piece of wood to which it is made to adhere with a little melted wax."

When young queens emerge in natural conditions they help themselves to honey and pollen and feed on these until they are mated. If caged without any provision of food they are not fed by the bees and soon die of starvation. As their physical development is still incomplete it is important that the candy given to them should be sufficient, and approximate as nearly as possible to their natural food. It may be made in the following way:—

Take a small spoonful of good liquid honey known to have come from a healthy stock and heat it to make it thin. Pour it into a warm cup and gradually knead into it some icing sugar, or better still, some castor sugar which has been ground to a fine powder. Add to the

PLATE XXXIII

[*Photo: L.E.S.*

Newly-emerged Queens in Nursery Cages

mixture as much pollen as can be taken from two or three cells and continue to add sugar and knead until the whole has the consistency of stiff putty, that is until it is not sticky to the fingers. Keep it in an air tight jar until needed for use.

A shallow frame of dimensions given on page 235, containing a battery of nursery cages $2\frac{1}{2} \times 2\frac{1}{4}$, each containing a queen-cell, is shown in Pl. XXXIII. All the queens but one have emerged. Several of them can be seen, the others hiding behind their vacated cells.

The period of imprisonment of the young queens must be as short as possible. They worry themselves in trying to escape and are receiving only an imperfect diet. Moreover the younger they are the more easily are they introduced to the nuclei or stocks into which they are to go.

Jay Smith (77, p. 52) noted some inferiority in queens from nursery cages. When caged for three days they mated later than others and some were physically defective. He observed also that the caged queens eagerly consumed the dried up remains of the royal jelly in the cells from which they had emerged. He therefore concluded that virgin queens need royal jelly which they must receive from the nurse-bees. There appears to be no direct evidence of this however and the writer, noting that Jay Smith fed his caged queens on "nice candy made out of powdered sugar and honey" suggests that the deficiencies in his caged queens were more likely due to the absence of pollen from their diet.

These considerations lead to the conclusion that nursery cages are not wholly desirable and that it is always better for young queens to emerge amongst the bees in the natural way.

A surplus queen-cell is always acceptable to queenless and broodless bees which have been confined for about six hours. Their confinement may continue, subject to their being ventilated and provided with food, until the queen emerges.

CHAPTER XX

THE QUEEN-MATING APIARY

To obtain the maximum of success in the improvement of stock by selection a special apiary which is as remote as possible from all other bees is necessary. Reference has already been made to the convenience of having an apiary some distance from the home apiary in connection with the establishment of nuclei without loss of field bees. Whilst complete isolation is rarely possible on account of the distances through which both drones and queens may fly at their mating times even a comparatively small bee-less area is valuable, and the proportion of mis-matings is progressively reduced as the area is increased.

Isolation.

Gillet-Croix (26, p. *37*) who records his nine years' experience with a queen-mating apiary came to the conclusion that both queens and drones may fly as much as 3 Kilometres (approximately 2 miles) from their hives to mate, and that therefore there should be at least a distance of 5 kilometres (approximately 3 miles) between the mating apiary and the nearest outside stock of bees if only desired matings are to occur.

In accepting these figures we have to consider the possibility of fugitive swarms invading the neighbourhood and also of migratory drones which may sometimes take up their abode in a strange apiary. These risks however are relatively unimportant if the mating apiary is liberally provided with selected drones.

It is not possible to guarantee that any mainland district will remain free of unwanted bees for any length of time. The best way to ascertain if it is free at any one time, apart from personal enquiries, is to inspect fields and hedgerows when nectar is being gathered, e.g., from dandelion, fruit blossoms, white clover, blackberry blossom. If no worker bees are seen, or only a rare one, the district is suitable for a mating apiary.

Complete isolation is sometimes obtainable on small islands lying several miles off shore, or at high altitudes, but these places involve difficulties of travelling.

If we cannot secure complete isolation we make the best of partial isolation. The mating apiary shown in Pl. XXXIV has been used by the writer for queen-breeding experiments for some years. It is situate $1\frac{1}{4}$ miles from the nearest apiary and is bounded on the north and east by trees which prevent direct flight in those directions. Small fruit trees give some shade in the hottest part of the day. Generally about 70 per cent of the matings have appeared to be satisfactory but in two seasons the results were affected by the presence of bees in a hollow tree about half a mile distant. When these had been destroyed others came to take their place.

A smaller mating apiary in use during the past season, situated amongst orchard trees in a woody hamlet on an open moor and $1\frac{3}{4}$ miles from the nearest apiary gave generally satisfactory results. The aim was to breed for docility, both drones and young queens being the offspring of the same mother. All the young queens, about thirty in number, produced progeny showing varying degrees of docility as compared with average bees of the district, but none quite equalled the original stock in this respect. Presumably therefore all the queens had mated with the selected drones, but if plural mating involving drones from the next apiary had taken place it would not have been indicated by colour deviations for all were of the same dark colour.

Jay Smith (77, p. *113*) writes "If there are no bees nearer than a quarter of a mile, and if you have all pure drones in your own yard, my experience would prompt me to say that you will have very little mismating". Whilst appreciating the value of this statement however, it is obvious that we should aim to secure a much greater degree of isolation.

Gillet-Croix mating apiary.

Gillet-Croix (26, p. *38*) desiring to Italianise his apiary of black bees introduced two drone-bearing Italian stocks to his queen-mating apiary with the result that nearly all his queens mated with the Italian drones. He describes the situation of this apiary as follows:—

About 1,500 metres (nearly a mile) from numerous stocks of black bees, is a large clearing at the edge of a pine wood, the trees of which reached a height of about 30 feet. The hives were placed inside a copse of pines within the clearing, and in front of them was an opening —"a kind of chimney" between overhanging branches, about 20 feet long and 10 feet wide. Although spaces between the trees allowed the sun's rays to reach the hives at times the bees rarely passed through them but ascended by way of the "chimney" to reach the open. This fairway, as we should prefer to call it, would at times be crowded with drones so that young queens emerging for mating could hardly fail to meet them. The obstructing trees prevented direct flights abroad and would cause both drones and queens to reduce speed and circle round to mark the position of their homes,—circumstances conducive to the desired meetings.

Gillet-Croix relates how, being puzzled in one year by an unusual number of mismatings and having blamed what he considered the insufficient height of his trees, he subsequently found that a fixed-comb beekeeper (skeppist) had concealed some of his hives under a hedge quite near to the apiary in the hope that his queens would mate with the Italians. He was disappointed in this, lost swarms through neglect, and finally removed them.

From these experiences we may note the use of trees in the isolation of an apiary, and the desirability of inducing our neighbours to co-operate with us in our queen-rearing. The writer permits his neighbours to place healthy and droneless nuclei in his queen-mating apiary when they have queens to be mated.

Environment.

In summer all hives need shade during the hottest period of the day, and nuclei especially because of their small entrances. If no shade is provided the bees will sometimes be forced to cluster outside a hive because of the intolerable temperature within. A nucleus hive should therefore be placed in partial or complete shade. Direct incidence of the sun's heat is not necessary to induce young queens to fly. They will do this, when impelled by desire, at any temperature which is not less than moderate. The drones however are induced to fly freely by direct sunlight, shade being less essential for them with their large-hive entrances and ample ventilation. Gillet-Croix recommends that the drone-bearing stocks be placed so as to have full sunlight from 9 a.m. to 5 p.m.

The shade of a bush fruit tree is excellent for a nucleus and the tree assists the queen in marking her proper location. A small obstruction such as a brick near the alighting board, or a stick or a small branch with one end forced into the ground near the entrance helps a queen to distinguish her hive from a neighbouring one. This is a matter of some importance for young queens returning from their mating flights are sometimes lost through entering the wrong hives. Especially is this the case where exactly similar nucleus hives are placed near together in a line like a row of soldiers—as is sometimes depicted in illustrations. Nuclei should be placed as far apart as is convenient, and each, with its surroundings, should be open to a clear approach and appear to a returning queen as a distinctive feature of the landscape. If on account of limited space they must be placed near together they should face in different directions and have differently coloured alighting boards.

[Photo: L.E.S.

QUEEN-MATING APIARY

Drone bearing stocks should be distributed symmetrically amongst the nuclei,—e.g., if there are three, one should be at each end, and one in the middle of the apiary.

Stimulation of drones.

A means of securing the simultaneous flight of queens and selected drones is described by some writers. It is considered to be useful in apiaries containing other drones and rests on the assumption that a young queen reaches a state of sexual desire on or after her 6th day. Briefly it is as follows:—

When young queens are about five days old each nucleus is provided with about two dozen of the selected drones. It is then closed, provided with stimulating food (e.g., honey and water in equal proportions), and taken into a cellar for about 24 hours. At the end of that period, which should be between 5 and 6 o'clock in the evening, when normal drone flying has ceased, the nucleus is placed on its stand, facing the sun, and the entrance opened. The imprisoned and stimulated bees surge out, together with drones and queen, the latter taking what may well prove to be her nuptial flight. If she is seen to return without the usual evidence of having been mated (p. 92) the procedure is repeated at intervals until success is observed (17, II, p. *322*).

Maisonneuve (62, p. *32*) ascribes this method to a German named Siebert who recommended stimulating the drone-carrying stocks daily during the previous fourteen days with a rich food consisting of warm liquid honey, pollen, and white of egg, a mixture which, he stated, will keep for a long time in a well closed vessel. Helmberg, a Canadian, also cited by Maisonneuve (62, p. *33*) varied the procedure by preliminary copious feeding of the nuclei followed by their detention in a cellar for three days.

Causes of failures.

It is in the mating apiary that we meet our irremediable failures,—some from causes beyond our control.

In the normal course of things we expect a young queen to be laying freely when a fortnight old although we know that in unfavourable weather this is sometimes deferred until she is as much as a month or more old (p. 48). If she is present but not laying within a fortnight we leave her in the hope that she may yet become mated, and meanwhile keep the feeder replenished. If not laying at the end of a month we suspect she has failed to mate. Soon she may prove to be a drone-breeder (p. 97) in which case she is to be removed, some fresh worker brood given, and a new protected queen-cell inserted one or more days later.

Occasionally the first brood of a queen whose laying has been delayed contains a proportion of drone brood. This may be due to some ovarian imperfection, or what is more probable, to the presence of laying workers. If due to the latter cause the trouble soon disappears but if to the former it continues and the queen should be rejected.

If she has been missing for many days and food has been supplied, laying workers will probably appear (p. 152). The nucleus is then worthless as such and should be united to a stock by the newspaper method.

Young queens are frequently lost at the time of their wedding flights. They may lead off small swarms which sometimes return, but more often disappear. They may be snapped up by birds, beaten to the ground in sudden storms, or they may return to the wrong hives. Spiller in a recent communication to the writer reported the loss of young queens which became entangled in freshly made cobwebs. Such losses should be detected at the weekly inspection and fresh queen-cells given.

Virgin queens, or queens recently mated, display considerable nervousness when attempts are made to catch them. They are then apt to take flight and sometimes fail to return. If it is necessary to catch one it is safer to place a partially open match-box over her than to attempt to seize her by the fingers. She can then be liberated at the inside of a window, seized by the wings, and caged.

When robbing is prevalent weak nuclei are likely to be robbed and their bees and queens killed or starved to death. To guard against this they should not be examined or fed during the day-time but only at evening. Their hive entrances must be kept small and they should be strengthened if possible by emerging brood from other stocks.

Supposed losses by dragon-flies.

Amongst the most beautiful of the orders of insects are the Odonata or "dragon-flies". They prey on other insects, including bees, and are considered by some bee-keepers to be especially partial to queens and therefore responsible for losses in the queen-mating apiary. They frequent pools and marshy places, where their young stages,—eggs and nymphs—are spent on the mud or on submerged stems of water plants.

A dragon-fly's enormous compound eyes give it keen vision which enables it to pursue and capture its prey in mid-air, and its strong mandibles combined with a broad upper lip or labrum form a powerful trap by which it seizes and retains its prey (5, p. *36*).

Root (73, p. *217*) states that dragon-flies, or mosquito-hawks as they are called in America, are very destructive of bees in some of the Southern States and speaks of damage done by them in a large apiary on the Apalachicola river to the extent of a thousand dollars in four or five days.

Whether these flies have a predilection for queens is uncertain. The writer runs several small apiaries, some situated on low-lying damp land, one of the latter being used for queen-mating. Every August numbers of dragon-flies settle on the nucleus hives in the mating apiary but rarely is one seen in any of the other apiaries. Usually from one to four are seen on the roof of one hive where they remain motionless for a considerable time unless disturbed. They do not however distinguish between hives containing fertile and virgin queens. It may be that they are attracted to the place because queens are frequently in the air, but there appears to be no direct evidence that they have a predilection for them.

Removal of fertile queens.

Young fertilised queens should be allowed to remain in their nuclei until they have deposited eggs in two or three of the combs so that the strength of the nuclei may be maintained. If they are to be sold as "tested" they should remain until their first brood has emerged. If however the nuclei are needed for immediate further service the queens may be removed as soon as a number of eggs are seen and the beekeeper is satisfied, from the earliness and regularity of their deposition, that they will produce workers and not drones. The nuclei should be taken to the headquarters apiary where the queens can be conveniently removed and caged, and fresh queen-cells inserted in their places.

Queen introduction.

When young queens are mated and laying it becomes necessary to introduce them to full stocks. This can be done safely only when the conditions of a stock and the queen to be introduced are taken into account. Many fine queens are lost through failure to appreciate the circumstances and to adopt the appropriate methods of introduction. The most commonly used method is described on page 325, and directions are usually given in books on ordinary beekeeping. The whole subject however is an involved one and the reader who would study it thoroughly is referred to the writer's book on "The Introduction of Queen Bees".

Tested queens.

The word "tested" is ordinarily used in a restricted sense in respect of a young mated queen. It means simply that her first sealed brood has been observed to be that of workers bees and that therefore she has been fertilised. Until her first flat-capped worker brood appears there remains the possibility of her being unmated and a drone-breeder. A mated queen can usually be certified as "tested" within the first month of normal life.

She may be tested as to visible characteristics of her progeny about a fortnight after the first sealed brood is observed, that is, when her earliest progeny emerge from their cells.

A further period of three or four months must elapse before it will be possible to certify that she is prolific and her progeny good-tempered and industrious.

Nucleus hive records.

In any queen-rearing apiary it is essential that some system of records be used. It is fatal to rely on the memory, especially if many hives are kept. Examinations may be avoided and much time saved if the detailed state of each hive can be ascertained at a glance.

A card index is ideal, but the busy beekeeper has little time or inclination to write in the course of his practical work.

The writer has devised the record card illustrated in Pl. XXXV. This is printed on strong paper which is pressed on to wet paint applied to the cover of a nucleus hive. When this is dry the whole of the surface is varnished to preserve the card from the effects of the weather.

A short fine screw at the centre driven a little way into the cover holds four "clock" hands, cut from sheet tin or zinc and painted black, which can be moved around it and are of lengths to reach their respective circles.

Adjustment of the screw tightens the hands to prevent slipping.

The key is as follows:—

1st (inner) Circle
Number of the month.

4th Circle.
Days of the month.

2nd Circle.
State of nucleus.
H. Hungry (needs feeding).
R. Relieve (too strong).
S. Strengthen, (weak).

3rd Circle.
State with respect to queen.
O. Queenless.
C. *Queen-cell.*
V. Virgin queen.
E. Eggs seen.
F. Fertile queen.
X. Mismated queen.
L. Lost queen.
D. Drone-breeding queen.
W. Laying workers.

It takes about half a minute to adjust the hands when leaving a hive and even less to read what is to be done at the next visit.

The record shown in the illustration indicates that on the twenty-third of August the nucleus contained a fertile queen (F) and that it needed to be fed (H).

PLATE XXXV

[*Photo : L.E.S.*

RECORD CARD FOR NUCLEUS HIVE

CHAPTER XXI

FINDING, SECURING, AND INTRODUCING QUEENS

In the course of queen-rearing it is frequently necessary to find and remove a fertile queen. This is not always easy. We may open a hive and find her on the first comb we examine, or we may look over all the combs two or three times and fail to find her. Apart from pure luck, much depends on the circumstances of the stock and even more depends on our procedure.

Circumstances of the stock.

There is seldom any difficulty in finding the queen of a small stock or nucleus. This rarely needs preparation with smoke or other intimidant, and the combs can usually be examined without appreciable disturbance of the bees. The queen will be found on one of the few combs of brood, usually on that in which she is laying eggs.

The same remarks apply to a normal stock in spring or autumn when only a small proportion of the combs contain brood. Provided the bees have not been unduly alarmed only the combs containing brood need be searched.

Difficulties arise in the case of large stocks, which during and after swarming time are crowded with bees, including drones, and which may have brood on most or all of the combs in the brood nest. At this time the bees are less docile than in the spring or late autumn and they often become impatient of exposure and disturbance before all the combs have been examined. If they are repeatedly quelled with smoke they may become disorganised and the frightened queen may elude search by leaving the combs and hiding on the hive sides, on the floor, or in the hive entrance.

Essential conditions for success therefore are that the bees have not been irritated by preparation and that when examined they and the queen are quietly moving over the combs and mainly intent on their ordinary duties. These conditions are usually attained by moderate preliminary smoking at the hive entrance succeeded by an interval of two to three minutes during which the alarmed bees feed themselves with honey. When well fed they are more docile and more patient of manipulation. It is a common fault to omit to allow this interval between smoking and examining, the result being that the bees often begin to sting the operator as soon as the hive is opened. In the case of a stock known to be bad-tempered a comparatively heavy smoking across the floor, inserted from back and front, should be followed by a five-minute interval before the hive is opened.

Procedure.

(*a*) *To find the queen of a stock in a single brood chamber.* Insert a few strong puffs of smoke at the entrance, or preferably under the brood-nest which can be slightly raised from the floor for the purpose by the hive-tool. After an interval of two to three minutes remove and cover the supers, if any. These will contain an increased number of bees and there will be correspondingly fewer in the brood-nest. Whilst the supers and excluder are being removed, as well as subsequently, the least possible amount of smoke should be used and this only *horizontally* over the tops of the frames to keep the bees down when necessary, or along the lugs of the frames to clear the way for the fingers. In no case should smoke be driven downwards between the combs as this would drive the queen from her normal place.

If the ends of the frames are propolised, as is usual

in the autumn, gently and quickly loosen them with the hive tool before beginning to examine the combs.

Remove one comb from each end of the hive. These usually contain stores, and no brood, so that the queen is unlikely to be found on them. Look over them quickly and place them temporarily in a nucleus box, thus leaving room in the brood-box for manipulating the other combs. Remove and replace these, one at a time, working alternately from both sides towards the middle. The queen often retreats from the comb about to be examined because this is partly exposed to the light.

If the bees are not in an unduly disturbed state the queen is not likely to be found on combs containing only honey and pollen, or on combs of sealed brood. Little time therefore is necessary for the examination of these. She will usually be on a comb with very young brood and eggs, or with eggs and empty cells prepared for eggs.

When withdrawing a comb, and while it is still within the hive and nearly vertical, take a rapid glance at both sides. In this way one can often see the light coloured under-side of the queen's abdomen which makes her conspicuous as she moves about the comb.

Withdraw a comb, and with back to the strongest light, hold it at the distance of your best vision. Acquire the habit of looking over the face of the comb systematically, e.g., first round the periphery and then round again, working towards the centre. If you suspect that the queen may be hiding between the comb and the frame place a finger near the place where you suppose her to be; she will always walk away from it.

As the queen walks over the combs amongst undisturbed bees the latter turn their heads towards her and at the same time move back to give her unobstructed passage. Consequently there is a small area in her wake which for a few seconds is clear of bees. This area often serves to indicate the location of the queen. On a very crowded comb however the unoccupied area does not usually appear. The queen herself is distinguished by her long and tapered abdomen which is less covered by the wings than that of a worker; by her large thorax; and by her long light brown legs, and the light-coloured underside of her body.

Careful inspection of one side of a brood-comb requires from 10 to 20 seconds according to the density of the bees on it, or an average of 30 seconds per comb. If an equivalent amount of time is allowed for the withdrawal and replacing of a comb, a ten-comb stock should be thoroughly examined within 10 minutes. Usually the queen is found in a shorter time. An experienced beekeeper knows on what combs she is most likely to be and begins by examining these. He often finds her on the first comb removed.

If the purpose of the examination is merely to confirm the presence or age of the queen, she and her comb should be restored to the hive without unnecessary delay and the hive closed. If she is to be removed she should be captured as described on pp. 322–323.

(*b*) *To find the queen in a double brood-nest hive.*

When a double brood-nest is used the queen has the free range of 20 or more combs in two brood boxes, one placed above the other. To inspect the 20 combs absorbs much time and manipulation and tries the patience of both bees and beekeeper. It is usually sufficient to examine only one set of combs if the following procedure is followed:—

With the hive tool raise one edge of the lower brood box from the floor and inject several strong puffs of smoke across the lower edges of the combs. Lower the brood box and wait three minutes. It is almost certain that the queen, who always moves as far as possible from smoke, will have ascended to the upper brood box. To make reasonably sure of this insert more smoke at the entrance and wait another two minutes. At the end of this time the queen will have settled down on the brood in the upper box. Remove this box to another stand and examine it separately, or alternatively place an excluder temporarily between the two boxes to prevent the queen from running from one to the other, and examine the upper one *in situ*. The writer has used this method for many years and almost invariably finds the queen in the upper brood box.

Occasionally a queen may have to be found in a stock with a triple brood-nest, especially when only shallow combs are used. In this case smoke skilfully applied from above and below, with intervals during which the queen may move away from it, results in her finally being found in the middle box, which as a rule, is the only one which needs to be examined.

(*c*) *To find the queen in a vicious stock.*

Remove the whole stock temporarily to a new position at a distance of not less than a dozen yards. On the old stand place a temporary hive fitted with a few empty combs and one comb of brood taken from the vicious or any other stock. The flying bees will return to the old stand, and after one good day of flying weather, only young and inoffensive bees will be left in the parent stock. The queen can then be easily found and removed, the stock re-united on its original stand, and a new queen introduced.

It is hardly necessary to say that any comb under examination should be held over the opened hive lest the queen herself fall off. A comb which must be kept out of the hive for some time should be accommodated in a spare nucleus hive. It should never be set on the ground with its top bar leaning against the hive wall, as the queen, if present and unnoticed, may walk off into the grass and be lost.

(*d*) *To find the queen in a stock prone to agitation.*

The great thing in searching for the queen is to have the bees and queen in a state of comparative tranquillity.

Some stocks are always tranquil and rarely if ever need intimidation. Others settle down quietly after a short interval of time subsequent to being smoked. Others are prone to become demoralised when disturbed and exposed to the light. The demoralisation, negligible at first, increases progressively until a state of panic is reached. The bees run pell-mell over their combs, cluster at the lowest edges and even drop to the ground in lumps whilst being examined. Whilst they are in this state it is uncomfortable and almost useless to search for the queen. The stock should be closed for a few hours and then searched again. A light preparation by smoking, succeeded by a two minute interval, and followed by a reasonably quick examination—young brood combs first—will usually result in the discovery of the queen before the symptoms of panic develop. Alternatively the same measures as recommended above for a vicious stock may be taken. Such a stock should of course be requeened as soon as convenient,—hence the necessity of finding the queen in such a stock on at least one occasion.

Marked queens.

It is obviously helpful in queen-finding if a queen is marked, (p. 57). The mark she carries on her thorax not only makes her conspicuous but may serve as an indication of her age. Its presence or absence indicates whether her stock has swarmed or superseded its queen during a given season.

Tomes' queen detector.

During the past two or three years G. A. R. Tomes of West Wickham, Kent, has devised and perfected an apparatus by means of which the presence and location of a queen may be quickly determined.

Early in her life a double disc of metallic foil, enclosing within it a minute quantity (·5 μg. to 5 μg.) of a radium salt, is attached by an adhesive to the thorax of the queen. She carries this throughout her life, and so constantly functions as a miniature transmitting set.

A special form of ionisation discharge tube, which is held in the hand, and which is sensitive to the radio-active emanations from the radium, is connected with a loud speaker which derives its power from a simple vibration unit coupled to a two-volt accumulator.

The discharge tube serves as a detecting rod and when passed over the combs, or if necessary between them, causes a crackling sound to be emitted from the loud speaker when it comes within about 8 in. of the queen, the sound increasing in intensity as the rod comes nearer to the queen.

The device was originally intended by the inventor to give warning of the emergence of a swarm, the alarm being given as the queen reached the hive entrance across which was placed the detector tube. On the suggestion of S. W. Gadge it was modified to serve as a queen finder for which purpose it has proved to be efficient and a great time-saver.

It may be used

(1) To locate a queen.
(2) To ascertain that a queen is present, e.g., at the end of the season, when brood-rearing has ceased. In this case all that is necessary is to place the tube above the quilts and to listen to the response.
(3) To detect the emergence and return of virgin queens when taking mating flights.

In a note on the announcement of this invention Miss Betts (13, p. *68*) suggests that a possible effect of radio-active matter on the thorax of the queen might be interference with cell-division in the ovaries with consequent production of defective bees, such as gynandromorphs.

The question of mutations induced by radio-activity and their bearing on queen rearing is discussed on p. 132.

In a recent communication to the writer the inventor

states that after extensive use of the apparatus no harmful effects have been observed. The device is likely to appear on the market at an early date.

To catch and cage a fertile queen.

If the queen is to be removed and caged it is desirable to have a cage, provided with candy and partially opened, in such a position that it can be quickly reached by the left hand and so ready for immediate use. When the queen is found grasp one end of the frame firmly with the left hand, at the same time allowing the other end to rest on some support such as the hive wall, but not in such a position that the queen can escape back into the hive.

Wait until the queen is walking away from the hand, and at a convenient moment seize her by the wings. This must involve a deliberate and firm movement of the thumb and first finger towards the thorax, not the abdomen. A little pressure on the former will do no harm but any pressure on the latter may ruin the queen.

It is fear of this risk that prevents many beekeepers from being successful in capturing queens. They hesitate at the last moment, and the queen, touched but not seized, slips away frightened. Each further unsuccessful attempt increases her alarm until, if she is not too old, she takes flight. Should this unfortunately happen the operator should stand still and leave the hive open for a few minutes. The queen will usually return and join the bees in the hive. She should be sought again some hours later, or on another day.

Having secured the queen with the right hand, put the comb down. Take the cage in the left hand with the thumb covering the opening which is to admit the queen. Raise the thumb and present the queen, head first, at the opening. As soon as she has entered the cage close the opening with the thumb.

Attendant bees, to be inserted in the same way, are then taken one at a time from the comb. Bees which are obviously very old or very young should be avoided as not being suitable for feeding the queen. About 10 attendants are sufficient for a short period of confinement, 15 for the three days needed for introduction by cage, and about 20 to 25 if to be sent through the post. In cool weather not less than 20 bees should accompany the queen in any case.

Catching a queen with a match-box.

A young fertile queen is specially apt to take to flight when attempts are made to seize her with the fingers. It is therefore safer to capture her in a match-box as follows:—

Open a match-box to about three-quarters of its length and place the opened portion over the queen and what attendants happen to be near her. Gently close the box, keeping it close to the comb whilst doing so. Secure the box by a drawing pin through the side. Take it indoors and release the bees and queen at a closed window, and one by one capture them with the fingers and insert them into the prepared cage.

Finding a virgin queen.

It is sometimes necessary to remove from a hive a virgin queen which has failed to become mated and to substitute for her a fertile queen. At other times a virgin, accidentally bred in a nucleus, may have to be discarded. In this case we have no combs of brood and eggs to guide us so that the queen may be in any part of the hive. If brood has been recently inserted however the virgin queen will usually be found in its neighbourhood.

She is smaller than a fertile queen, is unnoticed by the workers, is active and elusive, and apt to take flight on small provocation. It is therefore sometimes difficult to find and capture her.

To find her, prepare the stock with smoke as for a fertile queen. Remove the two outside combs first, examine them cursorily and put them temporarily into a nucleus-box. Examine the remaining combs thoroughly taking one from each end alternately, so working towards the middle. She will be recognised, when she comes into view, by her abdomen which is less covered by the wings and more tapered than that of a worker.

If she is to be caged it is best to catch her by the match-box method and then to transfer her to a cage behind a closed window.

In the course of queen-rearing however we seldom have to find and cage a virgin queen. Wishing to avoid the risk of inducing robbing, a serious menace to nuclei in late summer, we abstain from unnecessary manipulations and judge of the presence and condition of a young queen by certain hive indications. Illingworth (41) describes these as follows:—

"My method is to open the nucleus without smoke at any time of the day, under reasonable conditions, and to look at the queen-cell. If it is open at the bottom, even if the trap-door is missing, I know the queen has emerged and is safe, and I close the nucleus immediately; if it is torn at the side I know she has been destroyed; if it is still closed the queen may be dead inside it, or the trap-door may have got closed again after she emerged, in which case a brief examination will reveal the fact; if the queen is not out I wait another day or two in case she may be all right. In the same way, to see if a queen has mated and started laying I use no smoke and just draw a comb out a little way; if no eggs are seen I close the nucleus at once and wait a few days, but if eggs are present I can then look safely for the queen and see what she is like."

Queen introduction.

In another work (79) the writer has dealt fully with caging, nutrition, and travelling of queen bees, and with their introduction to stocks under varying circumstances and by different methods.

Since queen introduction is a necessary sequel to queen-rearing, every breeder should have a comprehensive knowledge of its principles, not only for his own advantage but also for that of his customers and friends. Since consideration of this subject is beyond the scope of this work only two serviceable methods are given here—one for fertile and one for virgin queens. They are reliable provided the hive circumstances are favourable. They are extracted from the summary of methods given in Chapter XII of the writer's book on the subject.

Fertile queens. Ordinary caging method.

Place the cage, containing the queen with or without attendants, on its side, across the frames, and over the brood (if any), so as to expose the wire-cloth covering to the hive bees. About 48 hours later replenish the candy if necessary and expose this to the hive bees. A day or two later ascertain if the queen has left the cage but do not examine the combs for a week.

Virgin queens. Caging method.

Sealed brood only, with stores, should be present. Keep the stock queenless for at least one day. Provide the cage with candy containing a little pollen. If a "press-in" cage is used fix it over sealed brood, honey and pollen. After three days break down queen-cells and expose the candy of the cage to the hive bees, or, in the case of a "press-in" cage, liberate the queen.

CONCLUSION

EVERY beekeeper can make his contribution to the general improvement of the bees of the country. This he can do by:—

(1) breeding only from queens whose progeny show a good record of behaviour and productiveness.
(2) destroying, after they are tested, the queens of inferior stocks.

He should bear in mind that the retention of a single inferior stock from which drones are allowed to fly may nullify all his efforts, as well as those of his neighbours, to secure improvement by selection. It may in fact result in a general deterioration of the bees in a whole neighbourhood. For this reason queens discarded on account of poor quality should never be sold or given to other beekeepers.

The small-scale beekeeper will probably take advantage of one of the methods described in Chapters VIII and XI, and breeding only from his best stock, will carefully eliminate all queen-cells which are prematurely sealed. If he relies on natural swarming for his queen-cells he will not make use of those from his less desirable stocks. He will restrict the production of drones in his poorer stocks whilst allowing their natural or augmented production in his best stock. Since he may not be able to establish a separate mating apiary he should arrange to share one with his neighbours if possible. An isolated site provided with a carefully tended drone-bearing stock is all that is necessary. Virgin queens bred at home would be taken in droneless nuclei to the site and brought home when mated. For this purpose there should be one nucleus hive for each two stocks in an apiary. A nucleus stock, formed early in July should provide at least two fertile queens before the end of September, apart from failures and losses.

The possibilities of local co-operation in queen-breeding deserve the attention of the Beekeeping Associations. We can imagine the advantages which would accrue to the beekeepers of a village, for example, if they combined to maintain a small isolated mating apiary and bred only from the choicest of their queens.

Requeening should be annual except in the case of stocks showing superior qualities. The queens of these should be retained for a second full season.

The beekeeper who needs a larger but yet a moderate number of queens each year will need to use one of the more advanced methods of breeding described in Chapters XIV to XVII. He may well avoid the difficulties and risks of grafting and use the Barbeau method,—or better still what the present writer for convenience has distinguished as the "B. H. S." method. This is at once the easiest and safest of the artificial methods and is attended by the best results.

The queen-breeder who needs large numbers and a continuous supply of queens will have recourse to Doolittle's grafting method or to the Barbeau method, since the number of queen-cells which can be prepared by these at any one time is unlimited.

The breeder who desires some degree of uniformity in his stock should breed from one of the pure races, and should endeavour to establish a desirable strain by a course of in-breeding.

In-breeding with hybrids will lead, not to uniformity, but to disconcerting variation, probable deterioration, and a much greater need for the elimination of undesirables.

In any course of breeding for the improvement of bees it is not sufficient merely to breed queens and drones from the best stock. The ruthless elimination of the unfit is of equal importance. In the case of other domestic stock the unfit are not allowed to breed, but with bees this cannot certainly be prevented on the male side, except of course, by the use of instrumental insemination which is not yet commercially available.

LITERATURE CITED

1.	Abbott, A. F.	Why I Like Caucasian Bees. "The Bee World", XVI, 7, 84		1935
2.	Appeler, C. W.	Tremendous Growth Force. "Gleanings in Bee Culture," March, 1922	Medina, Ohio	1922
3.	Audibert, A.	Plus de Miel	Montfavet, d'Avignon, Vaucluse	1934
4.	Bassindale, R.	Occasional Appearances of a Worker or Queen Bee from Eggs laid by Worker Bees. "The Bee World," XXVII, 2, 12		1946
5.	Bastin, H.	British Insects	London	1917
6.	Bateson, W.	Mendel's Principles of Heredity	Cambridge	1930
7.	Benton, F.	The Honey Bee	Washington	1896
8.	Bertrand, E.	La Conduite du Rucher	Paris	1937
9.	Betts, A. D.	Practical Bee Anatomy	Benson, Oxon.	1923
10.	Betts, A. D.	Press Mirror. "The Bee World," XV., 9, 102		1934
11.	Betts, A. D.	Research Notes "The Bee World," XV, 5, 59 and 141		1934
12.	Betts, A. D.	Research Notes "The Bee World," XV, 12, 142		1934

13.	Betts, A. D.	Arena. "The Bee World," XXV, 9, 68		1944
14.	Betts, A. D.	Research Notes. "The Bee World," XXVII, 1, 5		1946
15.	Butler, C. G.	The Incidence and Distribution of Some Diseases of the Adult Honeybee (*Apis mellifera L.*) in England and Wales. "The Annals of Applied Biology," XXXII, 4, 344–351		1945
16.	Caird, K. F.	The Mating Flight of the Queen. "The Bee World," XVI, 9, 99–102		1935
17.	Cheshire, F.	Bees and Beekeeping	London	1888
18.	Columella, L. I. M.	De Re Rustica, IX, 2. Scriptores Rei Rusticae, (Gesner, M.)	Leipzig	1735
19.	Cowan, T. W.	The Honey Bee	London	1904
20.	Cowan, T. W.	The British Beekeeper's Guide Book	London	1913
21.	Crew, F. A. E. and Lamy, R.	The Genetics of the Budgerigar	London	1935
22.	Darlington, C. D.	Recent Advances in Cytology	London	1937
23.	Doolittle, G. M.	Scientic Queen-Rearing, 5th Edition	Sandpoint, Idaho	1909
24.	Fleischmann, A. and Zander, E.	Beiträge zur Naturgeschichte der Honigbiene. Bibliothek des Bienenwirtes, III	Klosterneuburg	1910
25.	Ford, E. B.	Mendelism and Evolution	London	1945
26.	Gillet-Croix, A.	Précis d'Apiculture et Sélection des Reines	Bertrix, France	1924

27. Haydak, M. H. Larval Food and Development of Castes in the Honeybee. Reprinted from the Journal of Economic Entomology, 36, 5 — S. Paul, Minnesota 1943

28. Haydak, M. H. The Food of the Honeybee. Paper No. 502, Miscellaneous Journal Series, Minnesota Agricultural Experiment Station — S. Paul, Minnesota 1943

29. Haydak, M. H. (et al.) Vitamin Content of Honeys. Paper 1963 Scientific Journal Series, Minnesota Agricultural Experiment Station — S. Paul, Minnesota 1942

30. Haydak, M. H. and Palmer, L. S. Royal Jelly and Bee Bread as Sources of Vitamins B_1, B_2, B_6, C, and Nicotinic and Pantothenic Acid. Paper No. 1947, Scientific Journal Series, Minnesota Agricultural Experiment Station — S. Paul, Minnesota 1942

31. Haydak, M. H. and Tanquary, M. C. Pollen and Pollen Substitutes in the Nutrition of the Honey Bee. Technical Bulletin, 160, University of Minnesota — S. Paul, Minnesota 1943

32. Heilbrunn, L. K. An Outline of General Physiology, 2nd Edition — Philadelphia and London 1943

33. Herrod-Hempsall, W. Beekeeping, New and Old — London 1930

34. Hollowell, E. A. The Culture of Winter Clover. Bulletin No. 119, U.S. Department of Agriculture. Cited by Root (73) — 1945

35. Hopkins, I.	Queen Rearing. "The Bee World", III, 7, 195		1921
36. Huber, F.	Observations on the Natural History of Bees	London	1841
37. Hudson, D. S.	A Method of Swarm Control Incorporating Requeening. "The Bee World", XXV, 4, 26		1944
38. Hutchinson, R.	Food and the Principles of Dietetics. 9th Edition. (Revised by Mottram, V. H., and Graham, G.)	London	1943
39. Hutchinson, W. Z.	Advanced Bee Culture. 3rd Edition	Flint, Michigan	1905
40. Illingworth, L.	The Camphor Treatment for Braula coeca. Circular No. 7. Cambridge and District Beekeepers' Association		1945
41. Illingworth, L.	Use of Extra Food Chambers in Queen-rearing. (Personal communication)		1945
42. Kitzes, Schuetts, and Elvehjem	Vitamins in Honey. Cited in "American Bee Journal", LXXXIII, 11, 420		1943
43. Komarov, P. M.	Influence of the Age of the Larvae and of the Number of Generations upon the Development of the Queen's Sex Organs. "The Bee World", XV, 7, 81		1934

44. Laidlaw, H. H., Junr. Artificial Insemination of Queen Bees. Journal of Morphology, LXXIV, 429–465 1944

45. Langstroth, L. L. The Hive and the Honey Bee Hamilton, Illinois 1911

46. Leuenberger, F. (Traduction Jaubert), Les Abeilles, Anatomie et Physiologie Paris 1929

47. Longfellow, H. W. The Song of Hiawatha, XXI, 197–202

48. Mackensen, O. The Occurrence of Parthenogenetic Females in some Strains of Honeybees. Journal of Economic Entomology, 36, 3, 465–467 U.S.A. 1943

49. Mackensen, O. Parthenogenetic Female Bees. (Personal communication) 1945

50. Manley, R. O. B. Questions. Marking Labels. "The Bee World", XIX, 6, 71 1938

51. Miller, C. C. Fifty Years Among the Bees Medina, Ohio 1911

52. Morrison, G. D. A Reason Why an Old Virgin Queen Cannot be Mated. "The Scottish Beekeeper", XX, 5, 72 1944

53. Morgan, T. H. The Physical Basis of Heredity Philadelphia, U.S.A. 1919

54. Morgan, T. H. The Theory of the Gene Newhaven, U.S.A. 1928

55.	Muzzati, G.	The Rearing of Queens from the Egg. " The Bee World ", XVI, 6, 61–65		1935
56.	Nelson, J. A.	The Embryology of the Honey Bee	Princeton, U.S.A.	1915
57.	Ophel, A. E.	The Kangaroo Island Project. " The Australian Bee Journal ", November, 1944. Cited in " The Scottish Beekeeper ", XXI, 8, 107		1945
58.	Pearson, P. B. and Burgin, C. J.	The Pantothenic Acid Content of Royal Jelly. Proceedings of the Society for Experimental Biology and Medicine, 48, 415–417		1941
59.	Pearson, P. B.	Pantothenic Acid Content of Pollen. Proceedings of the Society for Experimental Biology and Medicine, LI, 291–292		1942
60.	Pease, C. H.	A Queen Rearing Experiment. " Gleanings in Bee Culture ", LXXIV, 1, 10		1946
61.	Pellett, F. C.	History of American Beekeeping	Ames, Iowa	1938
62.	Perret-Maisonneuve, A.	L'Apiculture Intensive et l'Élevage des Reines 3me Édition	Paris	1926
63.	Phillips, E. F.	The Rearing of Queen Bees	Washington	1905
64.	Phillips, E. F.	Beekeeping	New York	1943
65.	Pratt, E. L.	Baby Nuclei	Swarthmore, Pa., U.S.A.	1904

66. Pratt, E. L.	Commercial Queen Rearing	Swarthmore, Pa., U.S.A.	1905
67. Pritchard, M. T.	Modern Queen Rearing	Medina, Ohio	1935
68. Rennie, J.	Acarine Disease Explained. Bee Disease Investigation, Memoir No. 6. North of Scotland College of Agriculture	Aberdeen	1922
69. Rennie, J. and Hervey, E. J.	Nosema Apis in Hive Bees. Bee Disease Investigation, Memoir No. 3. Reprinted from the "Scottish Journal of Agriculture," Vol. 2, No. 2	Aberdeen	1919
70. Roberts, W. C.	Multiple Mating of Queen Bees proved by Progeny and Flight Tests. "Gleanings in Bee Culture", 72, 6, 255		1944
71. Root, A. I. and E.R.	The ABC and XYZ of Bee Culture. 1929 Edition	Medina, Ohio	1929
72. Root, A. I. and E.R.	The ABC and XYZ of Bee Culture. 1940 Edition	Medina, Ohio	1940
73. Root, A. I. and E.R.	The ABC and XYZ of Bee Culture. 1945 Edition	Medina, Ohio	1945
74. Rosser, J. H.	Queen Rearing. "The Bee World", XV, 10, 111		1934
75. Sladen, F. W. L.	Queen-Rearing in England	London	1905
76. Sladen, F. W. L.	Queen-Rearing in England	London	1913

77.	Smith, Jay	Queen Rearing Simplified	Medina, Ohio	1923
78.	Snelgrove, L. E.	Swarming, Its Control and Prevention	Bleadon	1934
79.	Snelgrove, L. E.	The Introduction of Queen Bees	Bleadon	1940
80.	Snodgrass, R. E.	Anatomy and Physiology of the Honey Bee	New York	1925
81.	Speicher, B. R.	Oogenesis, Fertilisation, and early Cleavage in Habrobracon. Journal of Morphology, 59, 401–21		1936
82.	Spiller, J.	The Mating Flight. "The Bee World", XVI, 2, 21 and XVI, 11, 128		1935
83.	Spiller, J.	The Mating of Bees. "The Bee World", XXVI, 6, 46		1945
84.	Tarr, H. L. A.	Abridged Report of the Brood Disease Investigation for the year ending September 30th, 1937. Reprinted from "The Bee World", November, 1937		1937
85.	The Apis Club	The Diseases of Bees		1934
86.	Tinsley, J. T.	The Rearing of Queen Bees. Bulletin No. 145 West of Scotland Agriculture College	Glasgow	1944
87.	Townsend, G.F. and Lucas, C. C.	The Chemical Nature of Royal Jelly. Biochemical Journal, 34, 1155–62. Cited by Haydak (27)		1940
88.	Varro, M.	De Re Rustica, III, 16. Scriptores Rei Rusticae (Gesner, M.)	Leipzig	1735

89.	Vergil	P. Vergili Maronis Opera (Connington) Georgic IV, 95–99	London	1865
90.	Von Rhein, W.	Über die Entstehung des weiblichen Dimorphismus im Bienenstaate. Arch. f. Entwicklungsmechanik, 129, 4	Berlin	1933
91.	Wallis, L. F.	The Sex of Bees. Proceedings of the Central Association, British Beekeepers' Association	London	1945
92.	Watson, Lloyd, R.	Controlled Mating of Queen Bees	Hamilton, Illinois	1927
93.	Watson, Lloyd, R. and Whitney, R.	Drones from Laying Workers are Fertile. "American Bee Journal", LXXXV, 5, 155		1945
94.	Wedmore, E. B.	A Manual of Beekeeping. 2nd Edition	London	1945
95.	White, M. J. D.	Animal Cytology and Evolution	London	1945
96.	White, M. J. D.	The Chromosomes	London	1937
97.	Whyte, R.	Laying Workers. "The Bee World", IV, 3, 75		1922
98.	Zander, E.	Der Bau der Biene. Handbuch der Bienenkunde, III	Stuttgart	1911

INDEX